Process Plant Instrumentation

by

Samuel Simons

UHAI

Uhai Publishing, Inc.
1731 Helderberg Trail
Berne, NY 12023
Toll free 1-877-842-4782
www.uhaipublishing.com

NOTICE TO THE READER

This book is based on research and the author's experience in the chemical processing industry. Neither the publisher nor the author make any further claims as to the fitness of the products or procedures represented herein. The reader is cautioned to consider the safety hazards involved in the procedures and process contained in this book and to adopt such safety precautions as are appropriate. The publisher disclaims any responsibility to obtain information from the manufactures of products other than that provided by the manufactures and contained herein.

Cover Design:
Bi-County Graphics

Cover Photograph:
PhotoDisc

Illustrations by:
S. Simons/ Bi-County Graphics

ISBN: 1-930528-05-1

Printed in the United States of America

For more information, contact:
Uhai Publishing, Inc.
1731 Helderberg Trail
Berne, NY 12023
Toll free 1-877-842-4782
www.uhaipublishing.com

1 2 3 4 5 6 7 8 9 10 XXX 06 05 04 03 02

Preface

This course is designed for entry level process technicians in the refining and petrochemical industries. It is intended to provide the knowledge and skills required by process technicians to successfully perform their necessary job task in a safe and efficient manner. The learning objectives are supported by the introduction of many technically related terms, concepts, and explanations. Line illustrations, charts, graphs, and tables are included to facilitate the student's comprehension The material is presented as a system of progressive building block modules, with the information in each chapter providing the foundation of skills and knowledge required in progressive chapters.

***Chapter One** introduces, identifies, and explains the fundamentals of process control including; fundamental control-loop components, signal transmission, open-loop and closed-loop control, binary logic, interconnecting wiring diagrams, piping and instrument diagrams, and control schemes.*

***Chapter Two** introduces, defines, and explains the physical principles of the process parameters of pressure, temperature, level, and flow, and identifies and explains the primary elements used to detect and measure these parameters.*

***Chapter Three** identifies and explains the function of pneumatic and electronic transmitters and transducers, physical displacement transducers, and the scaled-value outputs of common transducers. Electric and steam heat trace are also covered.*

***Chapter Four** explains the function, operation, and maintenance of chemistry analyzers. This chapter includes sections covering pH analyzers, oxygen analyzers, conductivity analyzers, infrared analyzers (spectrometers), and gas chromatography. Also included are a description of the distillation process and the application of gas chromatography in fractional control.*

***Chapter Five** defines and explains controller response characteristics and, identifies and explains the function of pneumatic, electronic, and digital controllers.*

***Chapter Six** identifies and explains the function, operation, and troubleshooting of final control devices including; air operated valves, motor operated valves, solenoid valves, and hydraulic operated valves.*

Chapter Seven *describes and explains the function, purpose, and application of emergency shutdown devices, redundant safeguards, fail-safe systems, interlocks, trips, and permissives.*

Chapter Eight *explains the function, operation, and troubleshooting of programmable logic controllers. Sections of Chapter Eight cover; PLC hardware, the binary numbering system, ladder diagrams, input/output addressing, and troubleshooting.*

Chapter Nine *provides an overview of distributive control systems. Subjects covered include; system hardware, system software, topology, and troubleshooting.*

Chapter Ten *encourages the student to transform the knowledge gained in previous chapters into practical skills. Detailed information is provided to guide the student through several realistic scenarios including; removing and restoring instruments and valves from/to service, performing automatic to manual and manual to automatic transfer functions, and field tuning control loops.*

Chapter Eleven *introduces and develops the application of troubleshooting skills. This chapter challenges the student with a realistic troubleshooting scenario where specific techniques and practices such as symptom recognition, symptom analysis, system manipulation, fault validation, half-splitting, and bracketing are developed to resolve a system fault.*

Chapter Twelve *reinforces safety, certainty, and professionalism in the work place. Topics covered include; safety awareness, communication skills, self-checking, peer checking, identification of error precursors, and risk reduction techniques.*

An Instructor's Guide is available to qualified instructor's whose students use this textbook. A PowerPoint® presentation is also available to help explain the content of the textbook in a classroom setting. Visit our web site at www.uhaipublishing.com for more information.

I dedicate this book to the memory of my father, Stud Simons.

I would like to express appreciation to my wife, Dora, for her patience and support.

Acknowledgements:
I would like to acknowledge the following people for their expert assistance and guidance in developing this text:

Louis V. Caserta, Alvin, TX
Lamar Garrett, Victoria, TX
Jane Medeiros, Hammond, LA
David D. Spann, Kenai, AK

Table of Contents

Chapter 1

Fundamentals of Control

OBJECTIVES

Upon completion of Chapter One the student will be able to:

- *List the fundamental components of a control loop.*
- *Describe the typical signal transmission path of a control loop.*
- *Convert equivalent linear scale values from pneumatic to current.*
- *Convert equivalent linear scale values from current to voltage.*
- *Differentiate between linear, square root, and log scales.*
- *State the advantages of square root and log scales.*
- *List the common control schemes.*
- *List the common engineering prefix for notation values.*
- *List the inputs to an elementary or line drawing.*
- *List the outputs from an elementary or line drawing.*
- *Define process, control, instrumentation, set-point, algorithm, measured variable, manipulated variable, controlled variable, feed-back, binary logic, truth tables, elementary/one line drawing, and piping and instrument diagram.*
- *Describe the purpose of process control instrumentation, a primary element, a transmitter, a controller, and a final-control element.*
- *Differentiate between open-loop and closed-loop control, bi-stable and analog devices, series and parallel circuits, feed-back, feed-forward, and cascade control.*
- *Identify, interpret, and analyze; piping and instrument diagrams and symbols for field and control-board*

instrumentation and interconnecting signal lines, elementary drawings and symbols for normally-open/closed contacts and relay coils, And gate symbol and truth table, Or gate symbol and truth table, Inverter gate and truth table and Latch gate symbol and truth table.
- *Convert engineering units and attach the appropriate prefix.*

1.0 Introduction

The term Process refers to a physical or chemical change of matter or conversion of energy. Control implies regulation or manipulation of the variables influencing a process in order to achieve a desired result. Instrumentation is a specific device or groups of related devices that measure, monitor, and/or control a process. The fundamental purpose of process control instrumentation is to measure, monitor, and control a process. For thousands of years processes were controlled completely by human manipulation. Skills such as firing clay pottery, baking bread, and smelting ore were passed on and handed down through the generations. A person or group of people directly observed the process, determined the control actions required, and manually manipulated the required variables to obtain a satisfactory product. Overall efficiency was poor and product quality was completely subjective. These early manual processes were often trial and error, and the results, based entirely on the personal skill of the artisan, were inconsistent. Modern processes are, for the most part, completely automated, and require minimal human intervention. Efficiency is high and product quality is rigidly controlled. Results are predictable and repeatable. The predominant human role in process production today is to monitor and maintain automated control systems.

1.1 Fundamental Components of a Control Loop

The four fundamental components of a control loop are the primary element, transmitter, controller, and final control element. Figure 1.1-1, demonstrates the function and association of the basic elements of a control loop.

Primary elements detect and convert energy from the process variable into a quantitative value suitable for measurement. The measured variable is the process parameter monitored by the primary element. Transmitters receive and interpret the measurement value from the primary element, convert it into a standardized signal of representative value, and transmit the standardized signal to the controller.

There are several scales relating transmitter/transducer signals to board readings. The most common are linear, square root, and logarithmic. Scales convert actual process values into

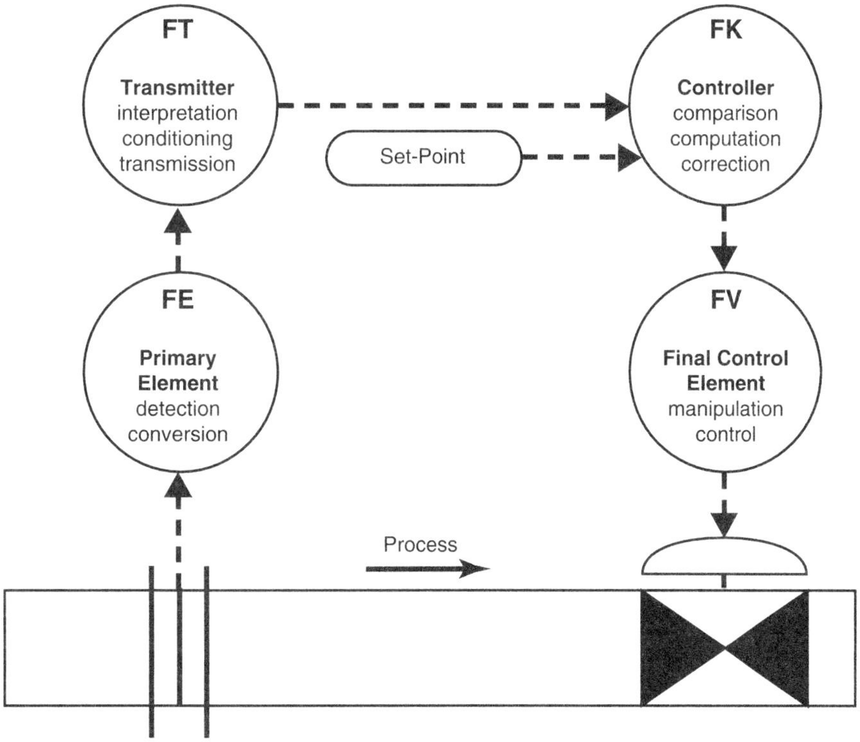

Figure 1.1-1 *Block Diagram of Simple Control Loop*

proportional analog transmission signals suitable for transmission, and back into indications of the actual process value for indication and recording purposes, see Figure 1.1-2. The primary element originally converts process energy into a detectable value for the transmitter. The transmitter, controller, and associated loop instrumentation processes and transmits the data as a standardized signal, such as 3–15 psig, 1–5 volts DC, or 4–20 mADC. The data is translated back into the original representative process value at the final indicating or recording point. The most common scale is the linear or percent scale. Linear scales are graduated to indicate an equivalent percentage of process value in relation to percent of instrument signal. Square root scales are graduated to extract non-linear flow values from a linear instrument signal. Logarithmic scales are commonly used in analytic processes. The advantage of Log scales is the ability to maintain excellent resolution throughout an extreme range. Figure 1.1-3 compares several types of scales. Figure 1.1-4 details the input and output spans of a simple pressure loop. Figure 1.1-5 details the actual process value signal at 50 psig as it is conditioned and transmitted through the loop. The transmitter converts the 50 psig process value into an equivalent current value of 12 mADC. The controller receives 12 mADC from the transmitter and compares it to the set-point (50 psig). With no offset (error) the controller outputs 12 mADC to the I/P. The I/P converts the 12mADC current signal into the pneumatic equivalent, 9 psig, and transmits it forward to the control valve. The control valve's available stroke is 0–2 inches from fully-closed position to fully-open position. Nine psig positions the valve at one-inch open. To convert linear scales calculate the span by subtracting the lower-range value from the upper-range value.

Process Value	Percent Span	psig	mADC	Volts DC
100 psig	100%	15	20	5
75 psig	75%	12	16	4
50 psig	50%	9	12	3
25 psig	25%	6	8	2
0 psig	0%	3	4	1

Process Value	Percent Span	psig	mADC	Volts DC
1000 psig	100%	15	20	5
750 psig	75%	12	16	4
500 psig	50%	9	12	3
250 psig	25%	6	8	2
0 psig	0%	3	4	1

Process Value	Percent Span	psig	mADC	Volts DC
220° F	100%	15	20	5
170° F	75%	12	16	4
120° F	50%	9	12	3
70° F	25%	6	8	2
20° F	0%	3	4	1

Figure 1.1-2 *Linear Equivalent Scale Values*

A temperature transmitter with a range of 20–220°F has a span of (220 (-)20) = 200°F. Divide the span by 100 to determine the number of units span equal to one percent (200 ÷ 100) = 2. In this instance two degrees of temperature is equivalent to one percent of the span. To determine the temperature value at 50% span, multiply the units per percent by the

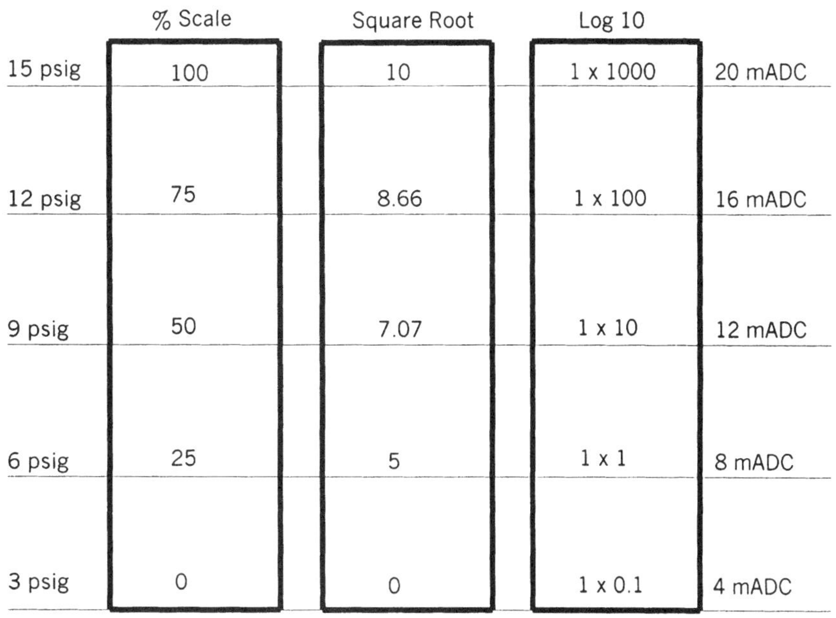

Figure 1.1-3 *Percent Square, Square-Root and Log Scales*

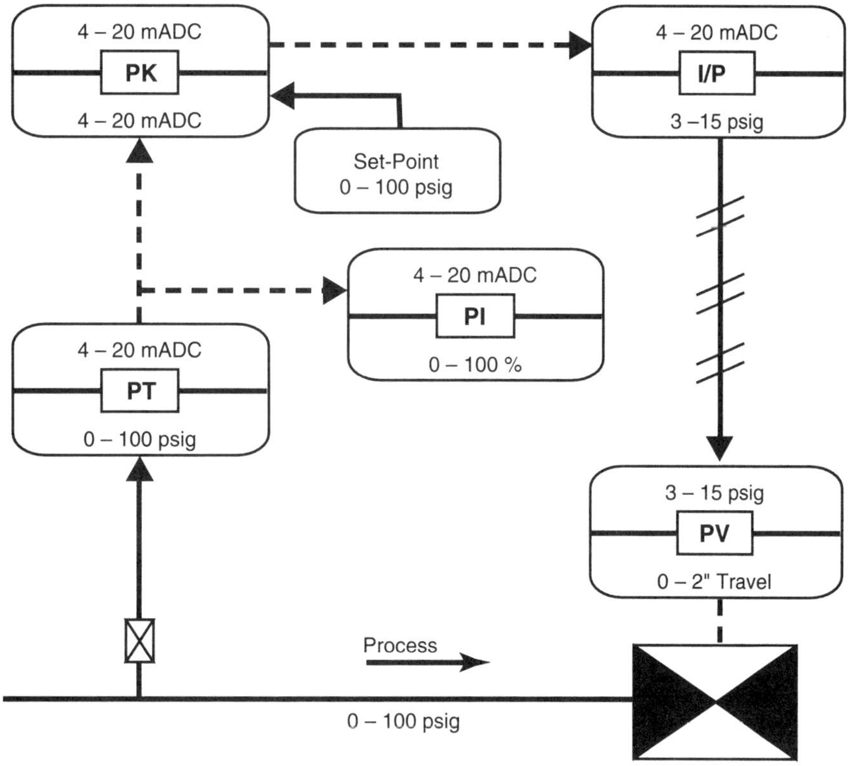

Figure 1.1-4 *Signal Conditioning and Transmission Path*

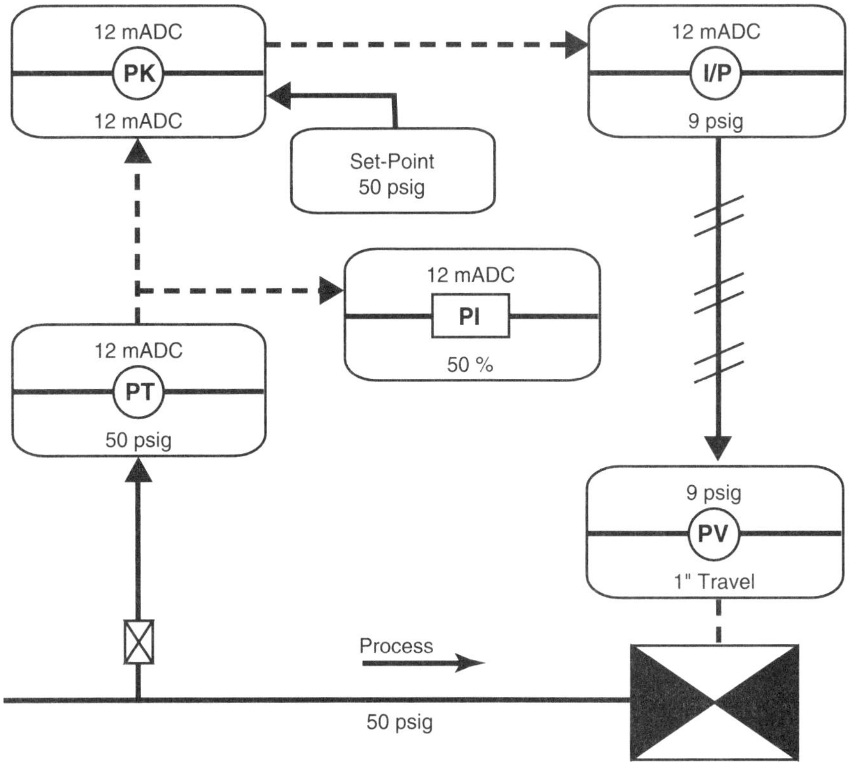

Figure 1.1-5 *Signal Values at 50 psig Process Pressure*

Figure 1.1-6 *Annunciator Panel*

Figure 1.1-7 *Two-Pen Recorder*

desired percent (2 × 50 = 100) and add the product to the lower range value (100 + 20 = 120). Fifty percent of 20–200°F is 120°F. This value is converted into the pneumatic equivalent by obtaining the 50% value of the 3–15 psig pneumatic span in the same fashion. Subtract the lower range value from the upper range value to determine the span (15 (-)3 = 12). Divide the span by 100 to obtain the units of value per percent of span (12 ÷ 100 = 0.12). Multiply the units per percent by the percent desired (0.12 × 50 = 6) and add the product to the lower range value (6 + 3 = 9). Fifty percent of 3–15 psig is 9 psig. Therefore 9 psig is proportional to 120°F. The same process is used to transpose all linear proportional scales. It is important to note that all signal ranges have a lower range value that is greater than zero. This provides for a "live" zero condition. A process value of zero percent will always generate some signal value, such as 3 psig, 4 mADC, or 1volt. A signal value below the signal's lower range value indicates a fault condition.

At this point the signal is conditioned sufficiently for indication, recording, or alarms. The controller receives the transmitter signal. Controllers, operating automatically, compare the signal to the desired value (set-point) and apply preset algorithms to create a correction signal. The set-point is the input variable that the controller recognizes as the desired control value. An algorithm is a pre-established set of specific, logical, or mathematical functions or instructions applied to any given variable. The controller then transmits the correction signal to the final control element. Final control elements receive the correction signal (demand) from the controller and respond accordingly. Final control elements function to directly change the value of the manipulated variable. If the controller algorithm is correct, the measured variable returns to set-point and the loop is stabilized.

Figure 1.1-8 *Multi-Point Recorder*

The manipulated variable is the process parameter directly altered by the final control element. Typical manipulated variables include; flow, temperature, pressure, level, and composition. The controlled variable is the process, parameter, or product over which a desired or influenced result is anticipated. In a simple feed-back-control flow loop the measured, manipulated, and controlled variables are sometimes the same variable. In most feed-forward control applications the measured and manipulated variables are not directly associated with the controlled variable, influence over the controlled variable is incidental.

1.2 Open-Loop and Closed-Loop Control

A closed-loop control system is any control system that has a feed-back signal. Feed-back is the ability to monitor the results of a control action and determine if additional correction is required. An open-loop control system is a control system operating without feed-back. In Figure 1.2-1, the open-loop system attempts to maintain the liquid level in the tank by cycling the valve on and off at preset time intervals. If all variables remain constant, this is an effective control technique. The closed-loop system directly monitors the level in the tank and manipulates the valve as required to keep the level constant. Closed-loop systems have the ability to automatically correct for disturbances. Figure 1.2-2 shows a simple closed-loop feed-back pressure control application. The valve is preset to remain at mid-position when the downstream pressure is at the desired value. If the pressure drops, the valve opens to allow more flow and increase the downstream pressure. If the pressure exceeds the desired value, the valve closes to restrict flow and decrease downstream pressure.

Many basic concepts are similar in all applications of control. Control schemes include a measured variable, a set-point, and a manipulated or controlled variable. A familiar control

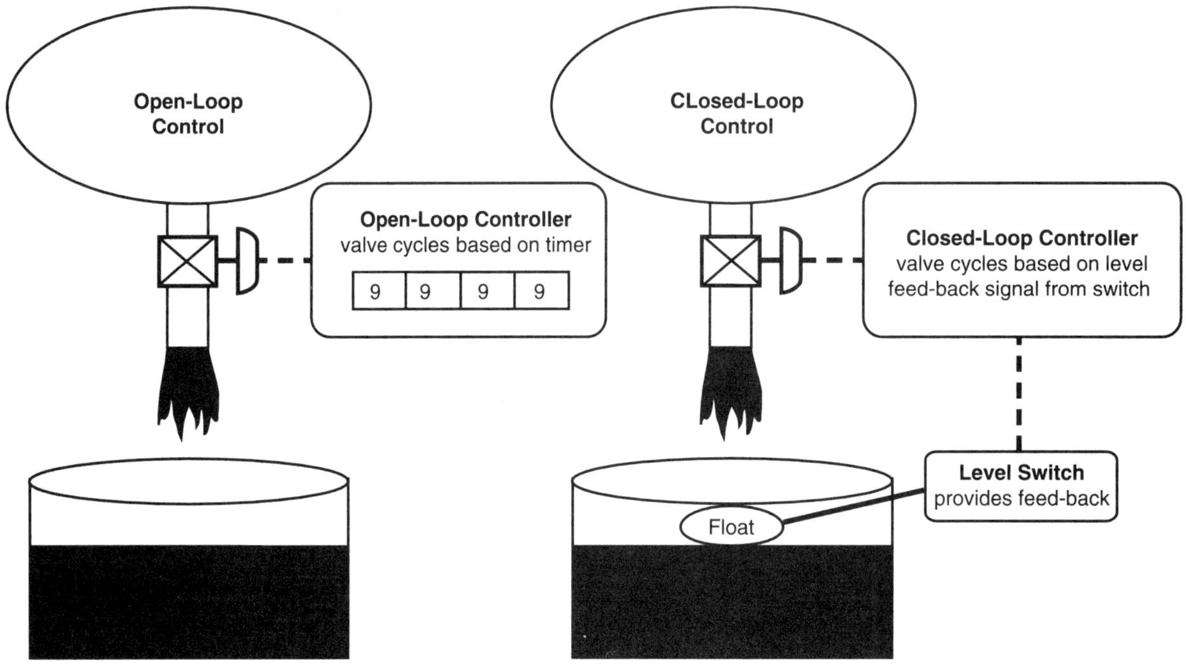

Figure 1.2-1 *Open-Loop/Closed-Loop Comparison*

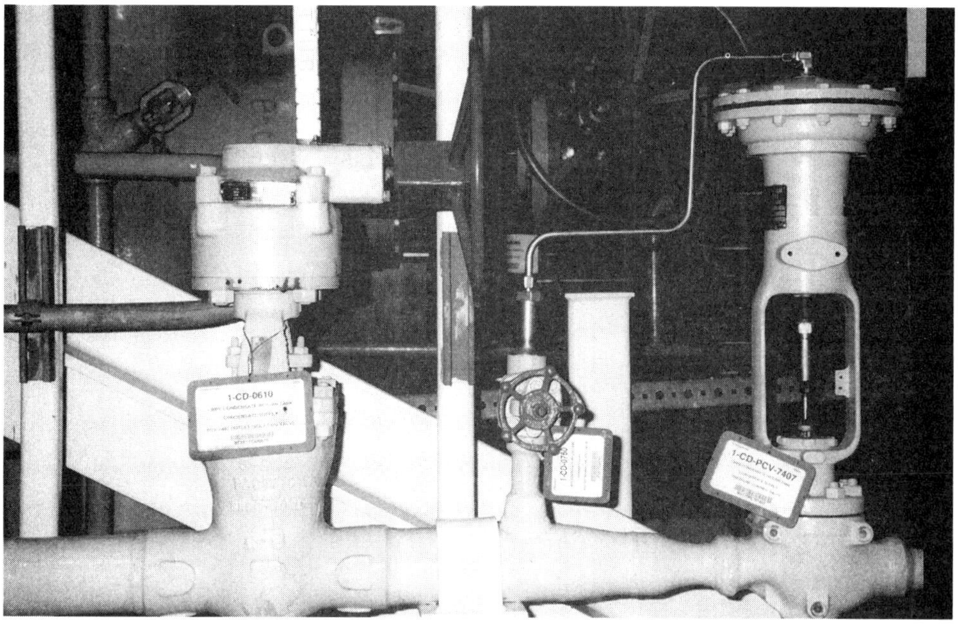

Figure 1.2-2 *Simple Pressure Control Feed-Back*

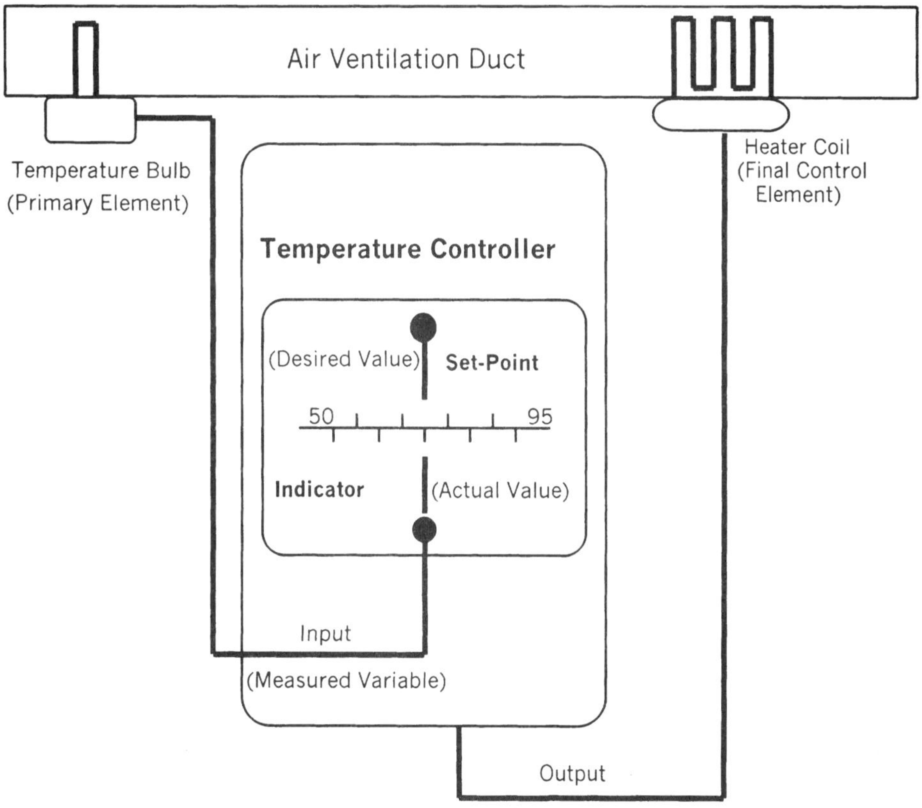

Figure 1.2-3 *Temperature Control Loop*

scheme, a temperature controller (thermostat) found in common central heating units, is depicted in Figure 1.2-3. The user selects a desired temperature, (set-point). The controller compares the measured variable (room temperature) to the set-point. The controlling unit detects any existing deviation and transmits a correction signal to the final controlling element (heater coil). The heating unit incorporates on-off (bi-stable) control. Bi-stable, on-off, control operates under one of two discrete states, ON or OFF. ON State is also referred to as (1) or TRUE, and the OFF State is referred to as (0) or FALSE. If room temperature is below set-point the heater turns on. The heater stays on until the room temperature matches set-point, and then it turns off. A low oil pressure warning light is also an example of a bi-stable device. If oil pressure is below set-point the light is on, if above set-point the light is off.

Another familiar example of closed-loop control is the cruise control system in a modern automobile. Again, the user selects a desired speed and the controlling system will attempt to maintain that speed. Cruise controls use analog control. Analog devices are capable of receiving, manipulating, and transmitting an infinitely variable set of values. They are limited only by the upper and lower limits of their design and their degree of resolution or sensitivity. If automobile speed drops below the set-point, the controller will incrementally attempt to accelerate the automobile back to match the preset set-point speed. If the speed error is small, the controller outputs a signal that demands the automobile to accelerate gently. If the speed error is large, the controller outputs a signal that demands the automobile to accelerate harder. An oil pressure gauge is another example of an analog device.

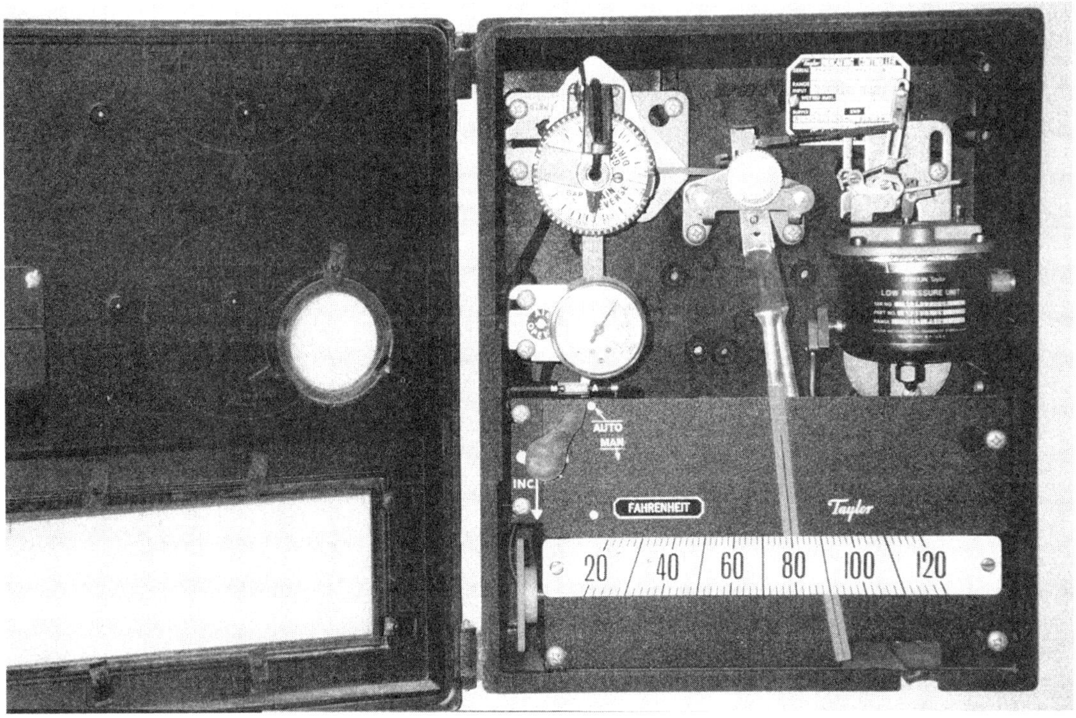

Figure 1.2-4 *Temperature Controller (Inside)*

Figure 1.2-5 *Temperature Controller (Outside)*

Application and design parameters determine the type of control selected. Analog control, although not generally used in residential heating, is acceptable for use in home heating units. But, if personal comfort is the main performance criteria, the additional cost and complexity add little to overall value. Bi-stable control is a simple and inexpensive alternative for analog automobile speed control units. The ride however, requires nerves of steel and considerable good fortune. The only control options available are accelerator full open and accelerator full closed.

1.3 Binary Logic

Binary logic is an application of mathematics that interprets, represents, and determines all possible values, states, or conditions based on a successive progression of TRUE/FALSE statements. Many process control systems are microprocessor based and operate in a digital format. Digital components process all information as a series of high and low states. A system of binary mathematics, Boolean algebra, is used to translate TRUE/FALSE logic into a string of ones (1) and zeros (0). TRUE is depicted as "one" and FALSE is depicted as "zero". A single digit of binary is referred to as a bit. Groups of eight bits are called a byte. The lowest order bit in a binary number, (2^0), is the "least significant bit". The highest order bit in a binary number, (2^n), is the "most significant bit".

Complex control schemes often contain a multitude of circuitry and components. It is impractical, if not impossible, to determine with certainty how each of the different components interrelates. Many times the control scheme is represented by set of logic diagrams that contain representative blocks, or gates, that map the functional operation of the system. There are only a few symbols used to represent the function of the system and the relationship between signals. The most common symbols are the AND/NAND gate, OR/NOR gate, NOT gate, and LATCH. The function of these gates is an application of Boolean algebra. Each gate performs based on a particular truth table for that gate. Truth tables are a chart of each possible output for all possible inputs to a logic gate.

Input	Output
0	1
1	0

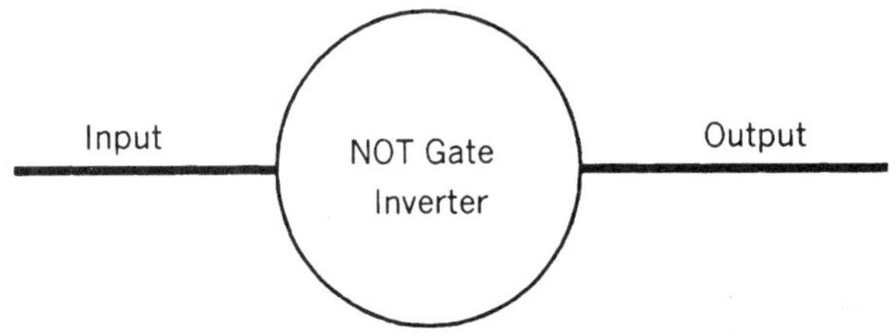

If the input is TRUE the output is FALSE.

If the input is FALSE the output is TRUE.

Figure 1.3-1 *Inverter Truth Table*

For any given input an INVERTER (NOT GATE) outputs the logical opposite as in Figure 1.3-1. The AND gate only outputs a TRUE (1) if all inputs are TRUE, Figures 1.3-2 and 1.3-3. A FALSE (0) applied to any input of an AND gate will result in an output of FALSE. A NAND gate is simply an AND gate with an INVERTER on the output.

Input A	Input B	Output AND	Output NAND
0	0	0	1
0	1	0	1
1	0	0	1
1	1	1	0

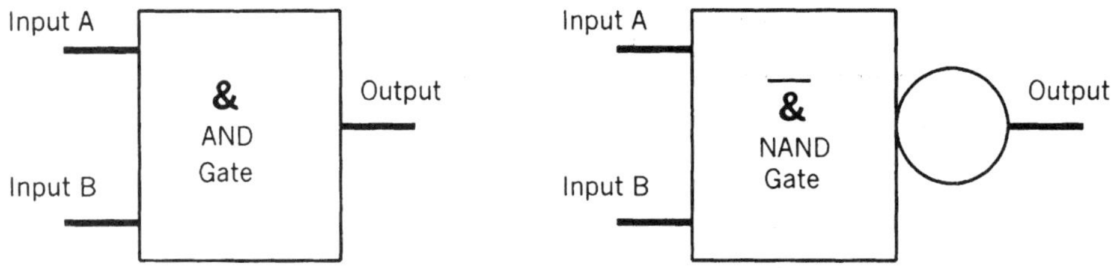

The outupt is true only if ALL inputs are true.

If ANY input is false the output is false.

The output is false only if ALL inputs are true.

If ANY input is false the output is true.

Figure 1.3-2 *AND/NAND Gate Truth Table*

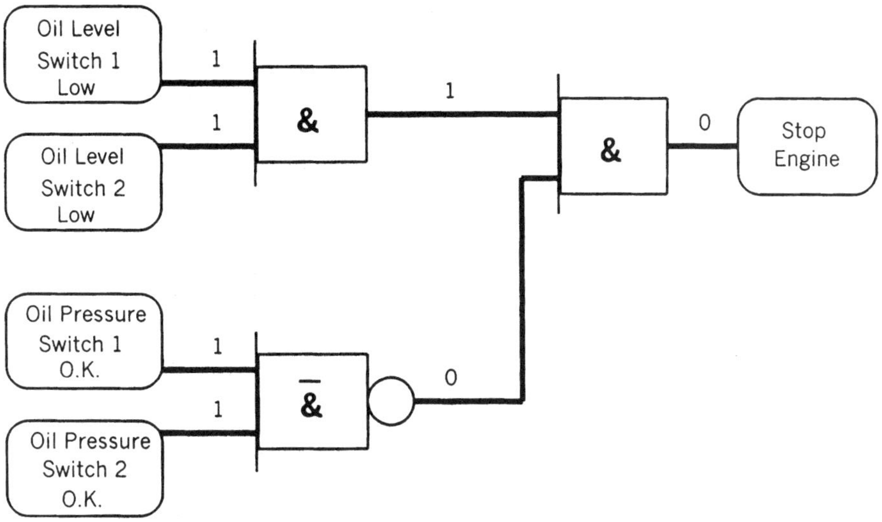

Oil level switch 1 AND 2 must indicate high to "enable" the stop engine gate. Once the stop engine gate is "enabled", oil pressure switch 1 OR 2 NOT O.K. will stop engine.

Figure 1.3-3 *AND/NAND Circuit*

The OR gate outputs a TRUE (1) if any of the inputs or TRUE, Figures 1.3-4 and 1.3-5. The OR gate outputs a FALSE (0) only if ALL inputs are FALSE. The NOR gate is simply an OR gate with an INVERTER on the output.

LATCH gates are used to catch and hold a logic condition, Figure 1.3-6. If both inputs are FALSE (0) the latch output remains CONSTANT at the last value. If the LATCH is RESET

Input A	Input B	Output OR	Output NOR
0	0	0	1
0	1	1	0
1	0	1	0
1	1	1	0

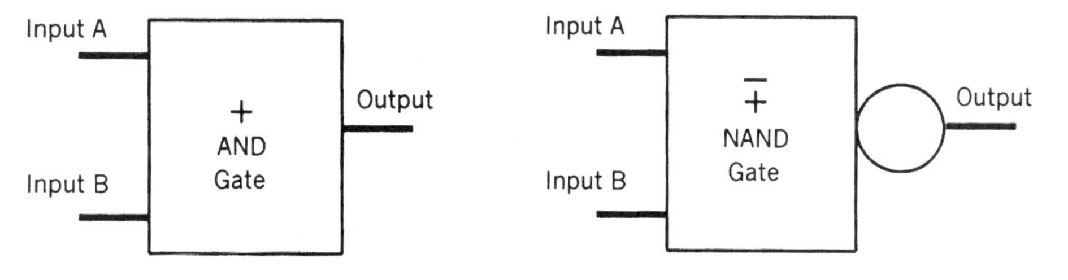

Figure 1.3-4 *OR/NOR Gate Truth Table*

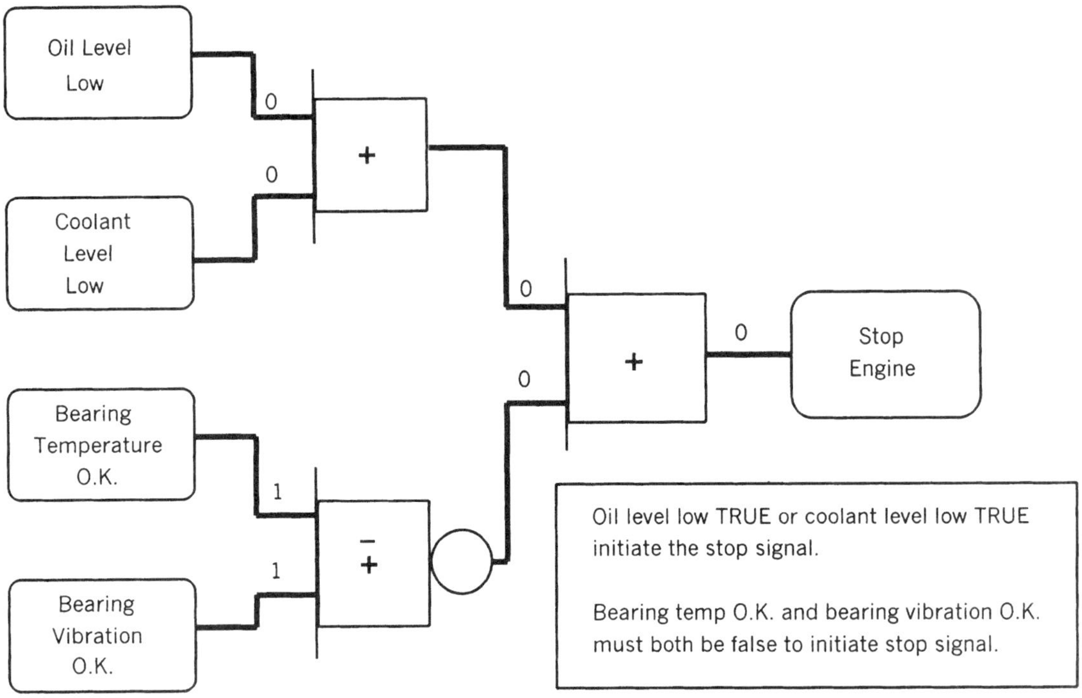

Figure 1.3-5 *OR/NOR Gate Circuit*

by logic TRUE on the RESET INPUT, the latch output, A, goes FALSE. The output will remain logic (0) until the SET input goes TRUE (1). If both SET and RESET are TRUE the LATCH will TOGGLE. The TRUE output turns FALSE and the FALSE output turns TRUE. A common application for a LATCH is to monitor start/stop pushbuttons as in Figure 1.3-7.

Input S	Input R	Output Q	Output $\overline{Q}$
0	0	No Change	No Change
0	1	0	1
1	0	1	0
1	1	Toggle	Toggle

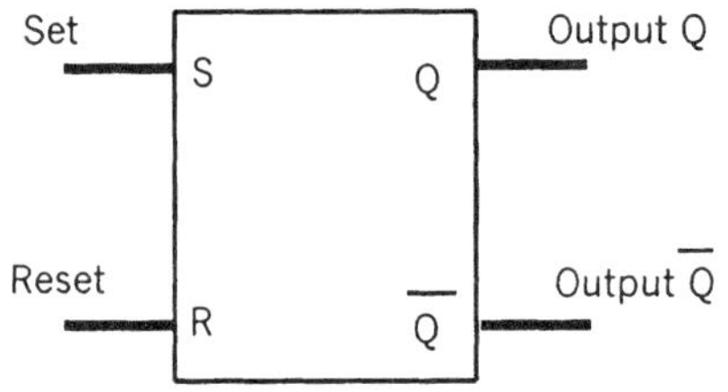

Figure 1.3-6 *Latch (Flip/Flop) Truth Table*

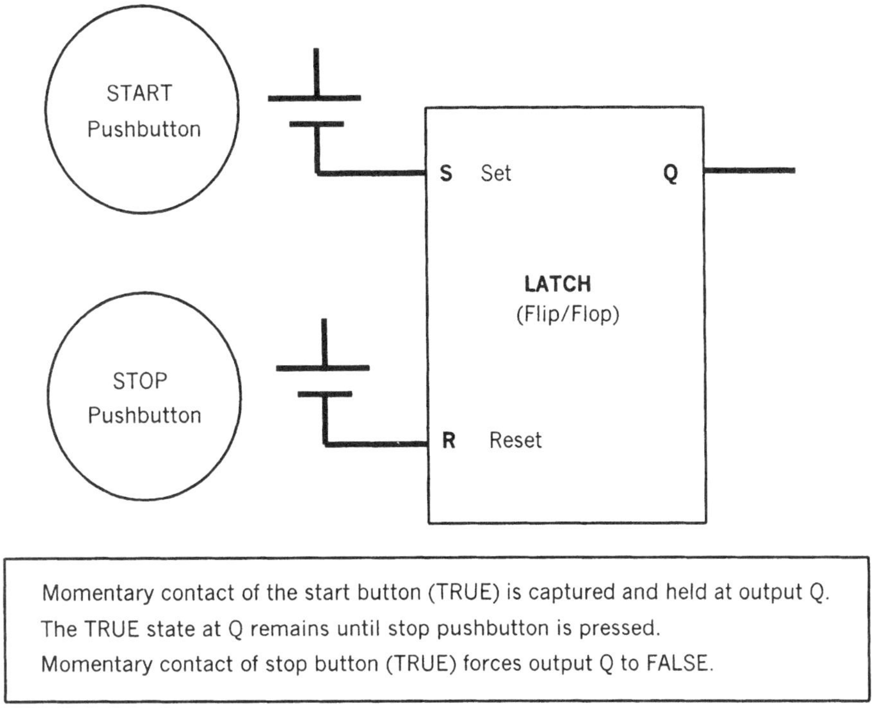

Figure 1.3-7 *Simple Latch Circuit*

1.4 Interconnecting Wiring Diagrams

Interconnecting wiring diagrams are drawings that depict control schemes as a system of series and parallel lines that represent the individual measurement and control elements of a process. Elementary and one-line drawings also illustrate electrical wiring schemes. These types of drawings portray relay, "hard-wired," logic and control applications. Commonly used symbols shown in Figure 1.4-1 are; contact normally open, contact normally closed, coil, and timer.

Elementary drawings begin the signal path at the load center or fuse box-supplying power to the particular loop. The power source is depicted as two parallel horizontal lines (runs) crossing the upper and lower sections of the drawing. The line (run) across the upper portion of the drawing is the input/power line. The lower line is the return line (run). Each individual circuit is represented as a vertical line, (branch, leg, or drop), connecting the upper and lower lines. Components along the same vertical line are wired in series. Components on vertically adjacent lines are wired in parallel.

Series circuits provide only a single path or route to conduct signal transmission or current flow as demonstrated in Figure 1.4-2. If the path is interrupted at any point all circuits on the path are de-energized. Signal transmission is only permitted if all contacts along the series path are conducting. Parallel circuits provide multiple paths for signal transmission

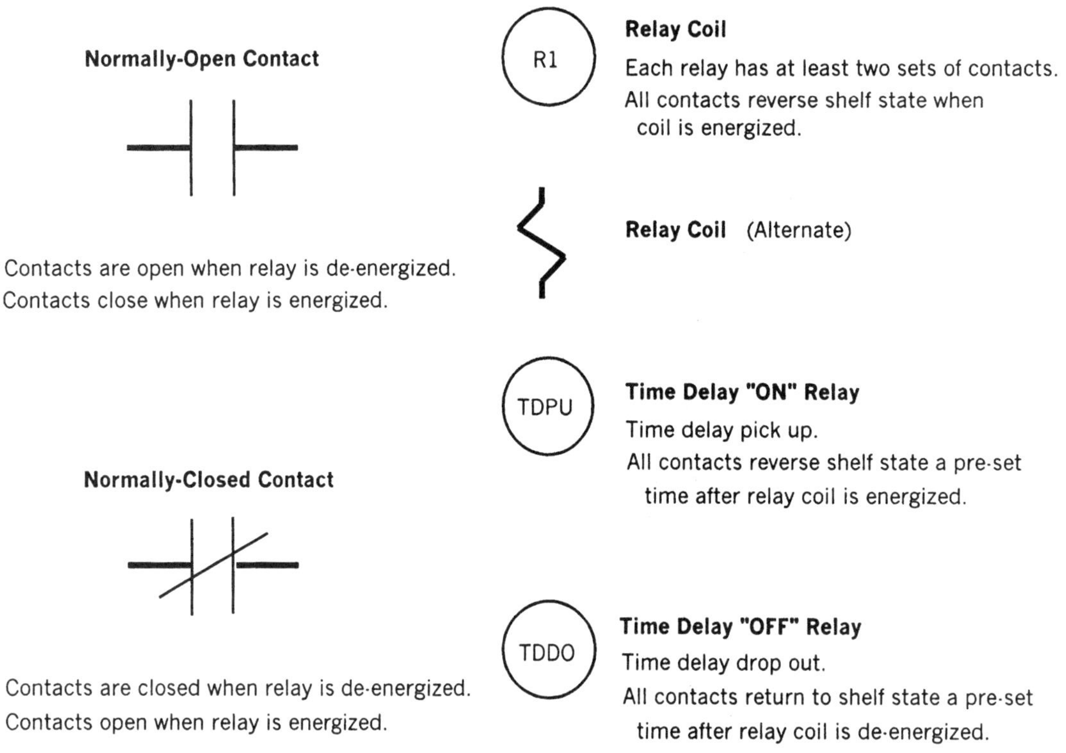

Figure 1.4-1 *Basic Relay/Elementary Logic Symbols*

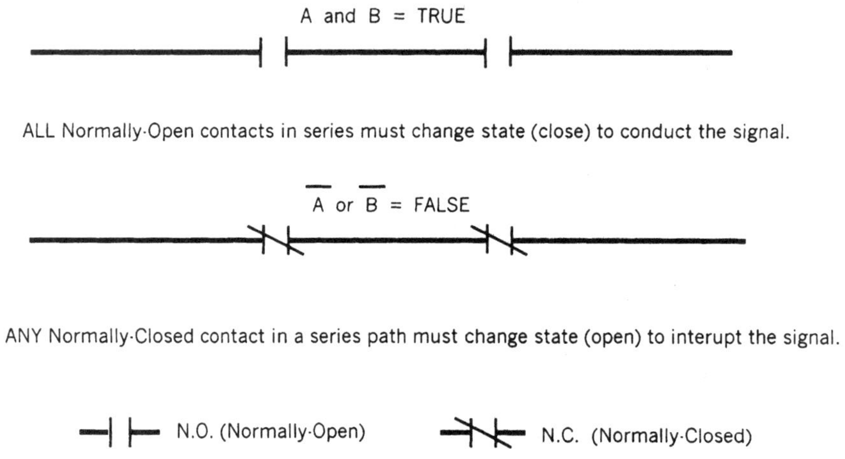

Figure 1.4-2 *Series Circuit*

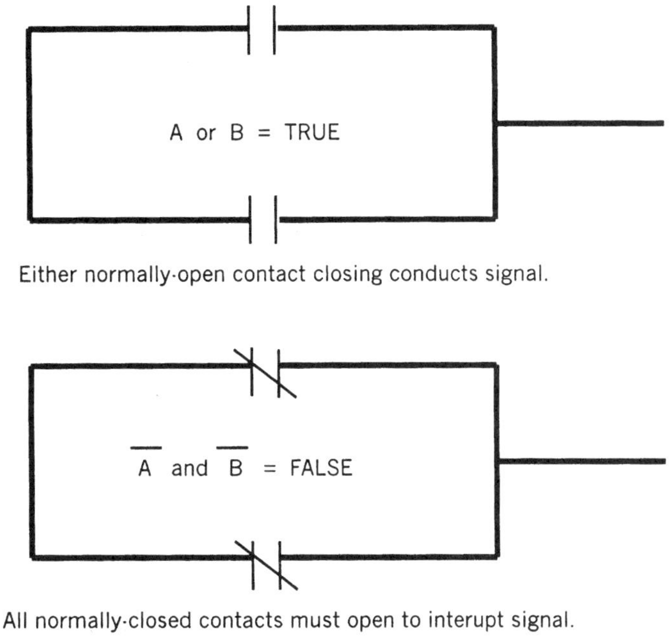

Figure 1.4-3 *Basic Parallel Circuits*

as demonstrated in Figure 1.4-3. If one path is interrupted the signal may continue (conduct) on an alternate or adjacent path.

Each branch in the circuit contains an input (switching device) and an output (coil/relay). The input is often a field contact from a pressure, temperature, level, flow, or valve position

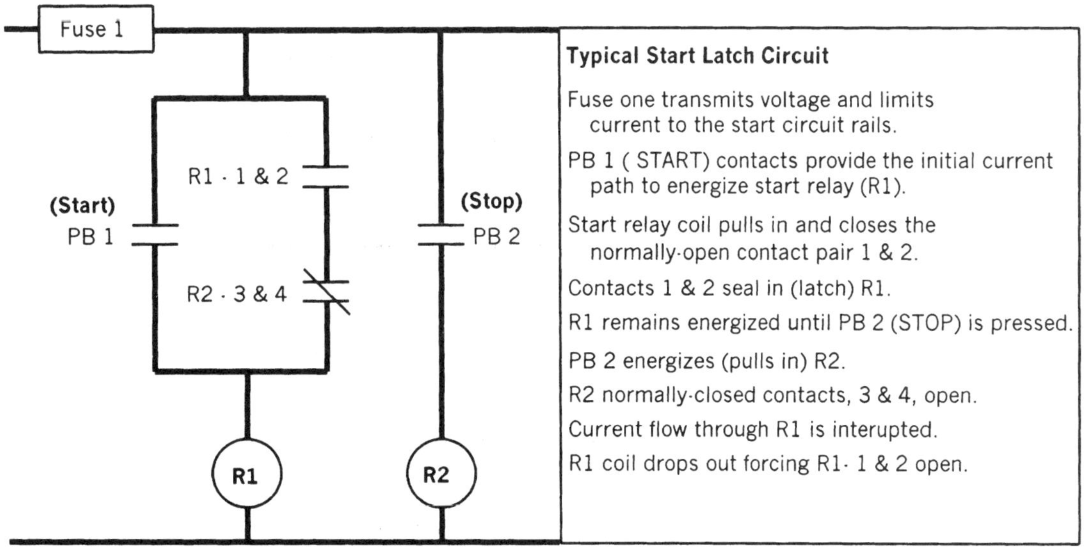

Figure 1.4-4 *Basic One Line Elementary Drawing*

switch. Pushbutton contacts, hand switches, relay contacts, and timers are also inputs. Outputs drive alarms, energize solenoids, start/stop motors and pumps, and provide logic states back to other circuits. The operational sequence of the circuit determines the selection or placement of input and output devices on the branches as in Figure 1.4-4.

1.5 Piping and Instrument Diagrams

Piping and instrument diagrams, P&ID's are illustrations that display the actual hardware components associated with the flow path of the process. Interpreting P&ID's is much like reading a road map. The process media progresses through piping as it is transferred from place to place. Process control component symbols are affixed at each representative monitoring point, and all processing devices are displayed. A multitude of symbols are used to diagram control loops. The first print in each set usually contains a legend depicting the

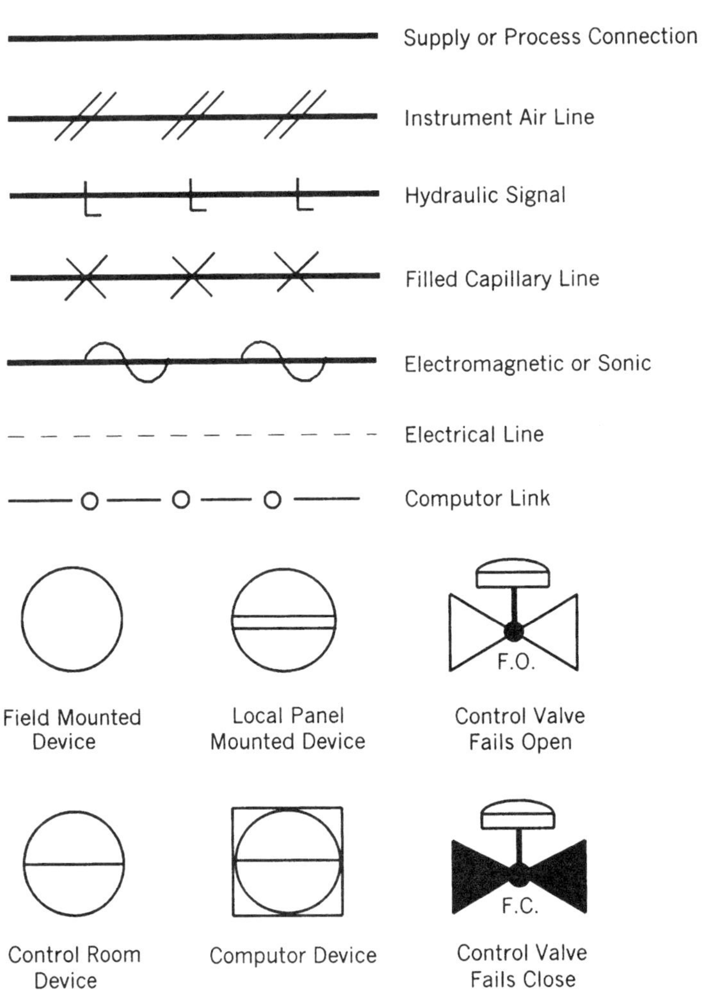

Figure 1.5-1 *Basic P&ID Instrument Symbols*

Letter	First Order Use	Succeeding Order Use
A	analysis	alarm
B	burner. flame	
C	user choice	controller
D	user choice	differential
E	voltage	element
F	flow	
G	user choice	gauge
H	hand	high
I	current	indicator
J	power	
K	time	control station
L	level	low
P	pressure	undefined
Q	quantity	totalize
R	radiation	recorder
S	speed/freq.	switch
T	temperature	transmitter
U	multivariable	multifunction
V	vibration	valve
W	weight	well
Y	event/state	solenoid/coil
Z	position	

Figure 1.5-2 *Instrument Function Abbreviations*

symbols used and their meaning. The American National Standards Institute and The Instrument Society of America publish the complete industry standard for instrument symbols and lettering abbreviations in the controlled document ANSI/ISA-S5.1.

Most instruments are represented as circles with the signal paths between associated instruments depicted as dashed lines, Figure 1.5-1, and the instrument type is identified by

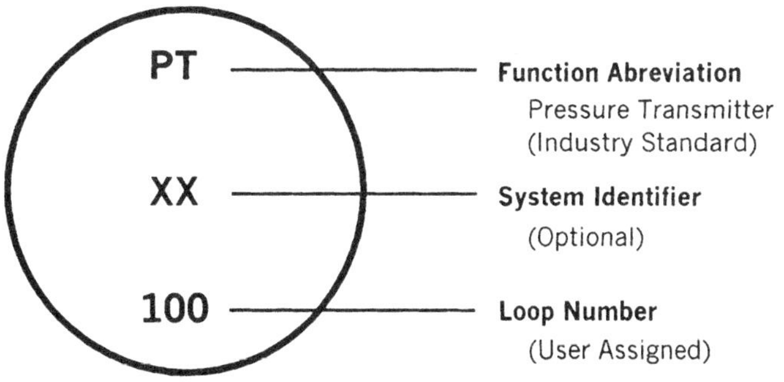

Figure 1.5-3 *Instrument Symbol Identification*

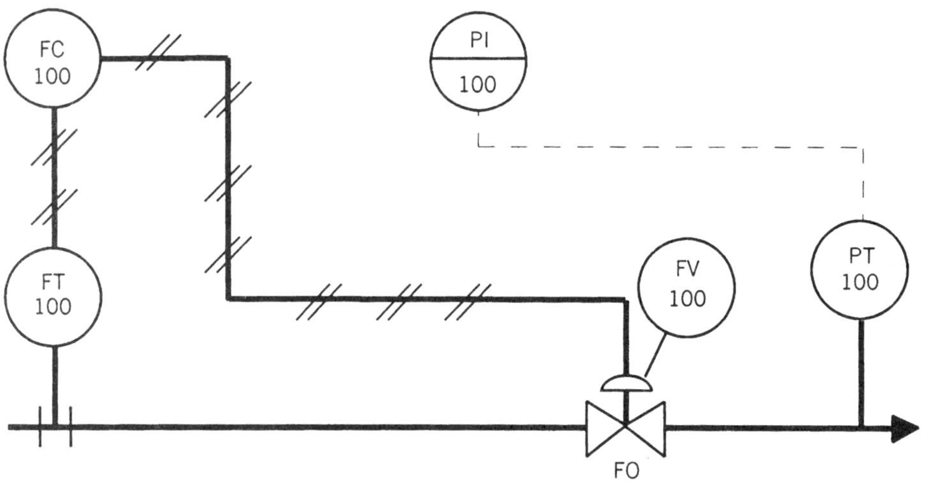

Figure 1.5-4 *Piping and Instrument Diagram*

an instrument function abbreviation, as listed in Figure 1.5-2. A numerical grouping hierarchy is also used to assist in clarification of component grouping and identification. Each instrument symbol should contain the function abbreviation, the process identification abbreviation, and the process control loop number, Figure 1.5-3. Figure 1.5-4 demonstrates all instruments monitoring and controlling a common variable sharing the assigned loop number.

1.6 Control Schemes

Common control schemes are; feed-back, feed-forward, and cascade. Control requirements of the process involved determine the type of control scheme employed. Selection parameters include; the degree of precision required, the inherent stability or volatility of the process, and the economic, environmental, or safety ramifications resulting from a lack of proper control. Each type of control has positive and negative attributes. Simpler control schemes are less expensive to install and maintain, but may not provide sufficient control for complex or sensitive processes. Complex control schemes are more expensive to install and maintain, but provide extremely stable and precise control.

Feed-back (reactive) control maintains the controlled variable at the desired value (set-point) by measuring the controlled variable, comparing the measured value to the set-point and manipulating the controlled variable until the set-point and measured value are the same. Figure 1.6-1 illustrates a typical example. Feed-back control cannot respond until after the controlled variable changes.

Feed-forward (pre-active) control measures one or more variables upstream of the controlled variable. A typical feed-forward control loop, such as in Figure 1.6-2, manipulates these vari-

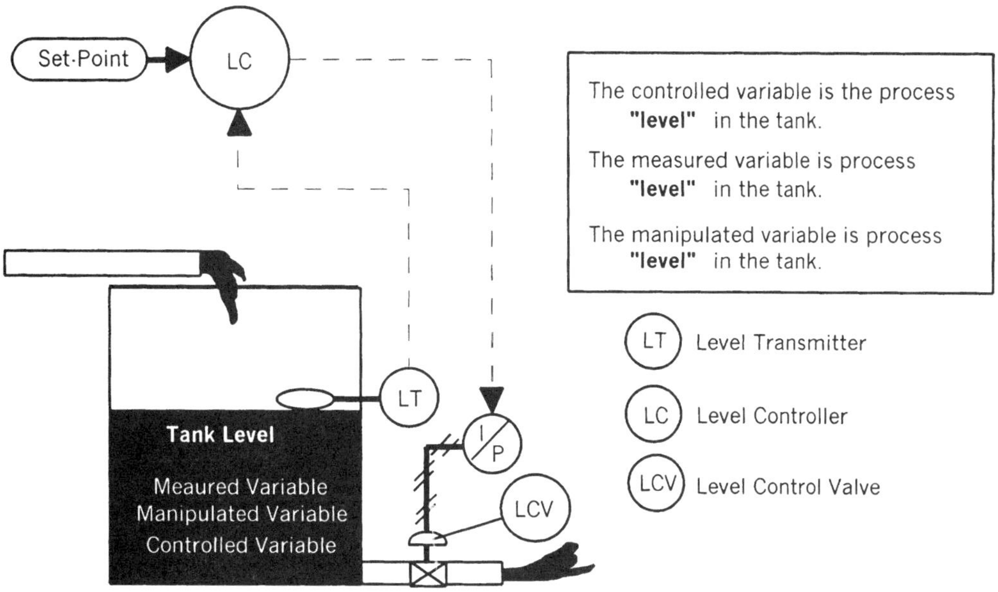

Figure 1.6-1 *Feed-Back Control*

ables, prior to their introduction to the process, to minimize disturbance to the controlled variable. All possible effects of upstream disturbances in the measured variables are calculated to pre-determine the appropriate degree of manipulation required to prevent an upset in the controlled variable. If designed correctly, feed-forward is a simple and stable control system. Any upsets or disturbances are measured and manipulated to prevent disruption of

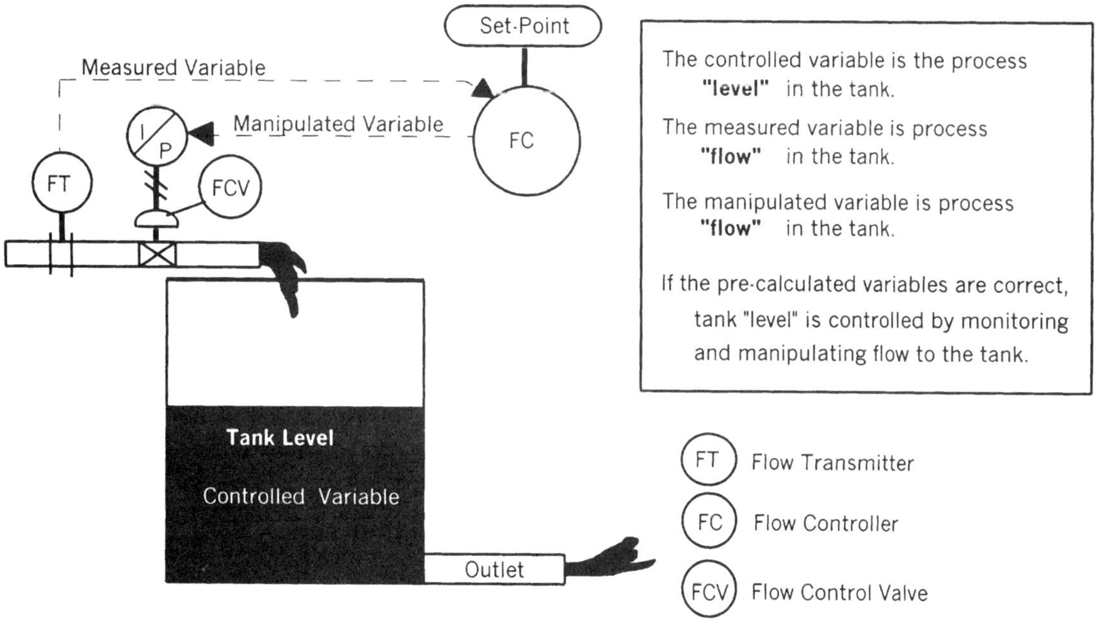

Figure 1.6-2 *Feed-Forward Control*

the controlled variable. Feed-forward control does not directly measure or manipulate the controlled variable, and is not capable of recognizing or responding to unanticipated deviations in the controlled variable.

Cascade (pre-active/re-active) control utilizes multiple controllers, with the output of one controller providing the set-point to another. A simple cascade loop is shown in Figure 1.6-3. The simplest concept of cascade control is an application that incorporates feed-forward and feed-back control in the same control loop. The output of the feed-forward controller establishes the set-point for the feed-back controller. In the illustrated example, the feed-forward section detects any disturbance upstream of the controlled variable. The disturbance forces an error signal in the flow controller. If flow increases, the controller outputs a new, lower, set-point to the level controller. In response to the new, lower set-point, the level controller interprets the negative error as an increased level. The level controller transmits a valve open demand to the final control element. This allows the feed-back section to begin reacting to the increase in flow before the level actually changes. The combined pre-act/react response of cascade control scheme results in an extremely stable system. The additional cost and complexity of this scheme limits applications to inherently volatile or demanding processes.

There are certain processes that benefit from alternate control schemes. The room heater discussed earlier in this chapter cycles continuously between on and off as it tries to maintain temperature precisely at set-point. Differential gap control would reduce equipment

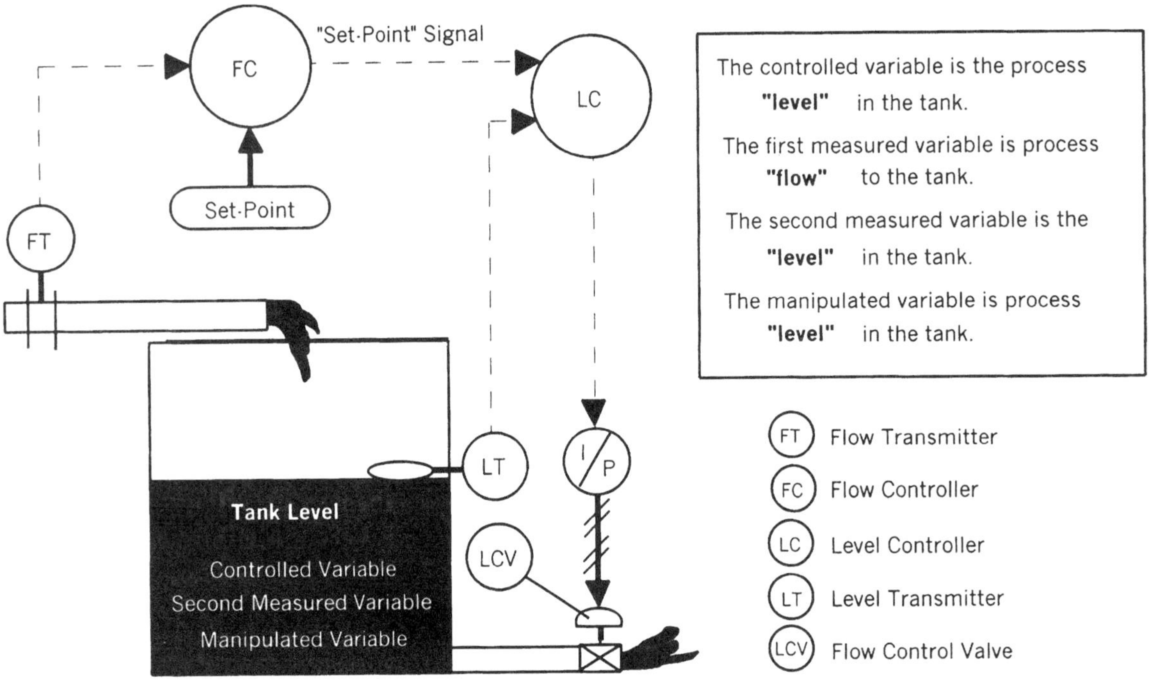

Figure 1.6-3 *Cascade Control*

wear and lower overall energy cost. Differential gap control attempts to maintain the process between a predetermined high and low value. This control demands the heater coil to energize when temperature drops significantly below set-point, and does not demand the coil to de-energize until temperature is at some value above set-point. The temperature range between the on and off demand is called dead band. While this is acceptable for a room temperature, some processs are quite intolerant of deviation and require more precise control. The automobile speed control unit discussed earlier in this chapter could benefit from multivariable control. Multivariable control accepts several different process variable inputs and computes an algorithm to produce the appropriate output demand. Monitoring engine vacuum and throttle position, in addition to speed, provides a more adaptive demand response. Ratio control is often used in mixing or chemical feed systems. Ratio control maintains a fixed ratio between two process variables. Complex processes often incorporate a master control. Master controls accept inputs from several subordinate controllers and output a total system demand to each controller under their control.

1.7 Engineering Notation

Manufacturing and process industries are large-scale operations. As such they produce massive quantities of product. On the other hand, quality, environmental, and efficiency concerns demand that extremely minute quantities be measured. Engineering notation simplifies the representation, communication, and documentation of this wide range of values. It annotates decimal place weights as groups of three. Figure 1.7-1 shows how each group of three places is assigned a specific prefix. A common transmission signal is "four to twenty milli-amps direct current". Engineering units allow representation of this value as 4–20 mADC. Without engineering units this value is 0.004–0.020 ADC. A common analyzer value is micro mhos. 10 micro mho is an easier term to use than the equivalent 0.000010 mhos. Electrical generators commonly produce 35 megawatts. Without notation this term is 35,000,000.00 watts.

Prefix	Symbol	Notation	Decimal Weight
Giga	G	$N \times 10^9$	N,000,000,000.00
Mega	M	$N \times 10^6$	N,000,000.00
Kilo	k	$N \times 10^3$	N,000.00
Centi	c	$N \times 10^{-2}$	0.0N
Milli	m	$N \times 10^{-3}$	0.00N
Micro	m	$N \times 10^{-6}$	0.00000N
Nano	n	$N \times 10^{-9}$	0.0000000N

Figure 1.7-1 *Engineering Notation*

Chapter One Summary:

- Process is a physical or chemical change of matter or conversion of energy.
- Control is the manipulation of variables influencing a process to achieve a desired result.
- Instrumentation is a device or group of related devices used to measure, monitor, and/or control a process. The purpose of process control instrumentation is to measure, monitor, and control a process.
- The four fundamental components of a control loop are the primary element, transmitter, controller, and final control element.
- The most common scales are linear, square root, and logarithmic. Scales convert analog transmission signals into indications of actual process value.
- The linear or, percent scale is graduated to indicate an equivalent percentage of process value in relation to percent of instrument signal.
- Square root scales are graduated to extract non-linear flow values from a linear instrument signal.
- Logarithmic scales are commonly used in some analytic processes. The advantage of log scales is the ability to maintain excellent resolution throughout an extreme range.
- The controlled variable is the parameter over which a desired or influenced result is anticipated.
- The measured variable is the parameter monitored by the primary element.
- The manipulated variable is the parameter directly altered by the final control element.
- The set-point is the desired control value.
- The primary element detects and converts energy from the process into a value suitable for measurement.
- The transmitter interprets the measurement from the primary element and transmits a standardized signal to the controller.
- The controller compares the actual value to the desired value (set-point) and applies preset algorithms to create a correction signal. An algorithm is a set of specific functions or instructions applied to a given variable.
- The final control element directly changes the value of the manipulated variable.
- Feed-back is the ability to monitor the results of a control action to determine if additional correction is required.
- Closed-loop control systems have feed-back. Open-loop control systems operate without feed-back.
- Bi-stable (on-off) control operates under one of two discrete states, on and off. On-off control is the simplest type of automatic control. Analog devices manipulate an infinitely variable set of values, limited only by the upper and lower limits of design and sensitivity.
- Binary logic evaluates conditions as a successive progression of TRUE/FALSE statements. TRUE is “1” and FALSE is “0”.
- Truth tables chart each possible output for all possible inputs to a logic gate.

- Interconnecting wiring, elementary, and one line drawings depict the function of individual measurement and control elements as a system of series and parallel lines.
- Series circuits provide a single path or route to conduct signal transmission or current flow. Parallel circuits provide multiple paths for signal transmission.
- Each branch in a circuit contains an input (switching device) and an output (coil/relay).
- Piping and instrument diagrams are illustrations that display the actual hardware components associated with the flow path of the process
- Each instrument symbol should contain the function abbreviation, the process identification abbreviation, and the process control loop number.
- The complete industry standard for instrument symbols and lettering abbreviations is established in the controlled document ANSI/ISA-S5.1.
- The common control schemes are: feed-back, feed-forward, and cascade.
- Feed-back (reactive) control measures the controlled variable, compares the measured value to set-point and manipulates the controlled variable until the set-point and measured value are the same. A deviation must first be sensed before corrective action occurs.
- Feed-forward (pre-active) control measures and manipulates one or more variables upstream of the controlled variable, prior to their introduction to the process, to minimize disturbances to the controlled variable.
- Cascade (pre-active/reactive) control utilizes multiple controllers with the feed-forward controller establishing the set-point and thereby changing the output of the feed-back controller.
- Engineering notation is used to simplify the representation, communication, and documentation of this wide range of values.

Chapter One Review Questions:

1. List the components of a control loop.
2. List the common control schemes.
3. List the common engineering prefixes for notation values.
4. List typical inputs on an elementary drawing.
5. List typical outputs on an elementary drawing.
6. Define the term process.
7. Define the term control.
8. Define the term instrumentation.
9. Define the term set-point.
10. Define the term algorithm.
11. Define the term measured variable.
12. Define the term manipulated variable.
13. Define the term controlled variable.
14. Define the term feed-back.
15. Define the term binary logic.
16. Define the term truth tables.
17. Describe feed-back control action.
18. Describe feed-forward control action.
19. Describe cascade control action.
20. State the purpose of process control instrumentation.
21. State the purpose of a primary element.
22. State the purpose of a transmitter.
23. State the purpose of a controller.
24. State the purpose of a final control element.
25. Differentiate between open-loop and closed-loop control.
26. Differentiate between bi-stable and analog devices.
27. Differentiate between on-off and continuous control.

27\. Differentiate between series and parallel circuits.

28\. Differentiate between feed-back, feed-forward, and cascade control.

29\. Illustrate the symbol for a flow transmitter on cooling water loop 100.

30\. Illustrate the symbol for a control board pressure indicator for cooling water loop 100.

31\. Illustrate the symbol for a fails-open flow control valve on cooling water loop 100.

32\. Illustrate the AND gate symbol and truth table.

33\. Illustrate the OR gate symbol and truth table.

34\. Illustrate a simple feed-back control loop.

35\. Convert 0.000010 volts to engineering notation.

36\. Convert 1200 ohms to engineering notation.

37\. Develop the logic diagram to stop the engine if oil pressure and oil temperature are both high, or if vibration is high.

38\. Develop the elementary diagram to latch "seal in" an engine over speed alarm relay.

Chapter 2

Primary Sensing Elements

OBJECTIVES

Upon completion of Chapter Two the student will be able to:

- *List and identify the primary sensing elements used to detect pressure.*
- *List and identify the primary sensing elements used to detect temperature.*
- *List and identify the primary sensing elements used to detect level.*
- *List and identify the primary sensing elements used to detect flow.*
- *Identify primary sensing element malfunctions.*
- *Define the following terms; primary sensing element, pressure, temperature, hydrostatic pressure, static pressure, differential pressure, vena contracta, head correction, specific gravity, conduction, convection, radiation, and absolute zero.*
- *Differentiate between flow rate and total flow.*
- *Differentiate between mass and volume measurement as it pertains to flow.*
- *Convert units of pressure.*
- *Convert units of temperature.*
- *Calculate level sensor span and range.*
- *Calculate specific gravity level corrections.*

2.0 Introduction

Primary sensing elements detect and convert energy from the process into a value suitable for measurement. They are used to measure a multitude of process variables; flow, level, pressure, temperature, composition, etc. Process reaction energy is detected in a multitude of physical forms. The characteristic of the event observed is determined by the critical reaction under control. The common method is to monitor the actual value of concern. As this is not always possible or feasible, secondary parameters are sometimes monitored, and the critical value is inferred or extrapolated. This chapter investigates process variables and the primary devices used to detect and convert these variables into quantitative values.

2.1 Pressure

Pressure is defined as force exerted per unit area, Figure 2.1-1. A seventeenth century mathematician named Pascal documented several fundamental principles that hold true for pressure. The first principle states that the pressure at any point in a fluid acts equally in all directions. The second principle states that a change of pressure exerted at any point in a body of fluid at rest is transmitted undiminished to every point in the fluid. Pressure exhibits these principles as it will always try to equalize throughout a system, Figure 2.1-2.

Scalar values of pressure require two terms. Practicality dictates common or accepted usages and conventions for combination of terms relating units of force to units of area. Common measured units of force are; Newton's, pounds, and ounces. Common measured units of area are square meter, square centimeter, square millimeter, square feet, and square inches. Conventional statements of pressure combine these common terms. Examples of these combinations are newtons/meter^2 and pounds/inches^2. Since only like

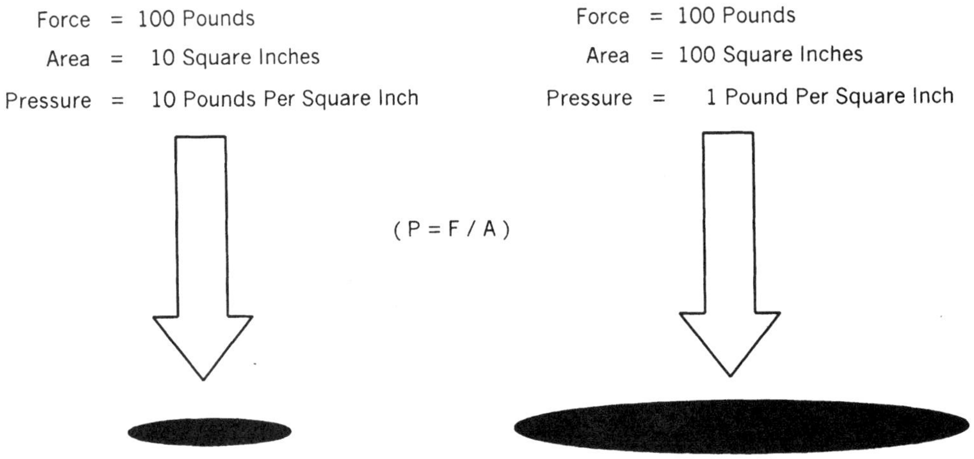

Figure 2.1-1 *Pressure Equals Force Divided by Area*

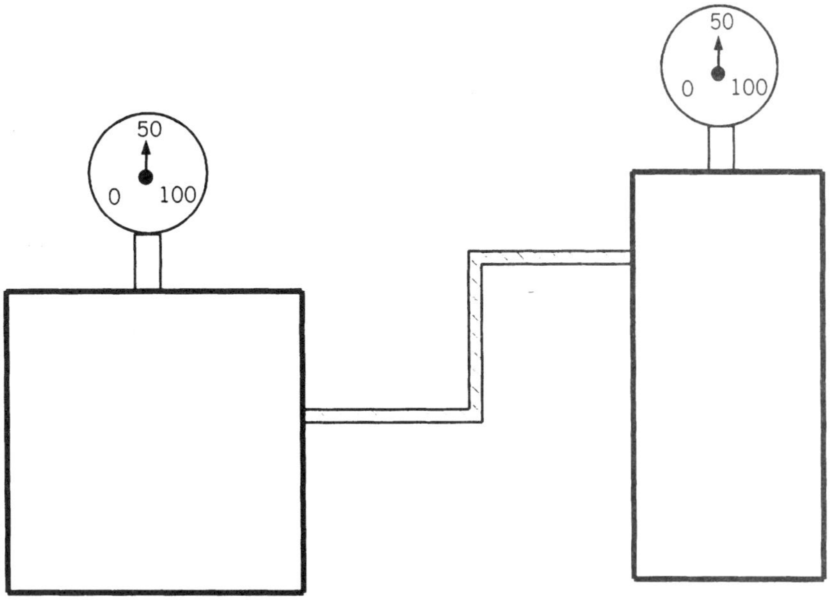

Figure 2.1-2 *Pressures Equalize Across Entire System*

terms are manipulated and compared, conversion tables are frequently used. Common types of pressure sensing elements include manometers, bourdon tubes, and compression displacement elements.

Other parameters affect pressure values. Pressure in an enclosed system is dependent on temperature and volume. A balance exists between these three parameters and that balance always remains constant. Robert Boyle (1627–1691) and J.A.C. Charles (1746–1823) developed the basis of this relationship for gases. Mathematically stated pV/T = 1 (maintains unity). If temperature increases, pressure and/or volume must also increase. If temperature decreases, pressure and/or volume must also decrease. If volume increases pressure and/or temperature must decrease. If volume decreases pressure and/or temperature must increase. The following example assumes certain ideal conditions and is simplified for clarity.

Example: If a 1000^3 inch vessel is filled to capacity, pressurized to 100 psia, sealed, and stabilized at 373 degrees Kelvin, what is the maximum temperature allowable if the vessel burst at 300 psia.

Note: *Temperature must be in Kelvin and pressures must be absolute.*
Units of absolute pressure and volume must be the same for both statements.

Solution: the initial condition is 100 psia × 1000^3/373 degrees Kelvin =1

The final condition is stated as 300 psia × 1000^3/N degrees = 1

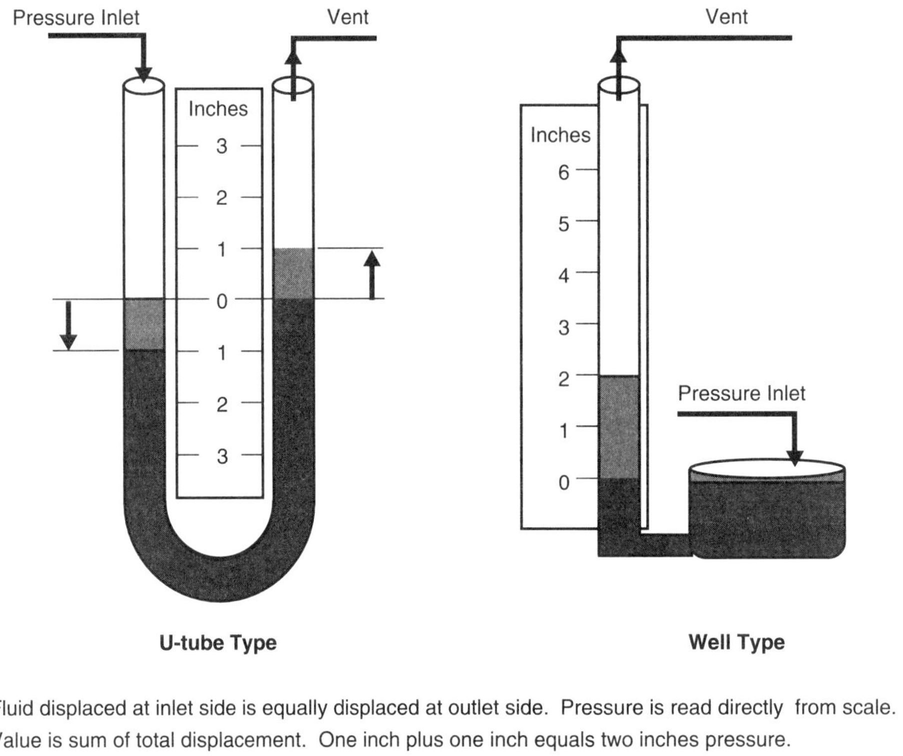

Figure 2.1-3 *Manometers*

Since the relationships MUST remain constant the initial condition is stated equal to the final condition.

If $100 \times 1000/373 = 268$, Then $300 \times 1000/N = 268$, Therefore $N = 1119$ degrees Kelvin.

Manometers are the simplest types of pressure detector. Manometers detect and convert a change in pressure into a linear vertical displacement of a known liquid. Evangelista Torricelli, who succeeded Galileo as professor of mathematics at the University of Florence, Italy, developed the first documented manometer. A manometer is basically a hollow transparent tube partially filled with pure water or mercury. As depicted in Figure 2.1-3, a ruled scale is affixed to the tube and the change of fluid level in the tube is read directly on the scale. The value indicated is equivalent to the amount of force required to lift the fluid the measured height. This type of pressure is called hydrostatic pressure. The fluid and scale used dictate the measurement units. If the fluid is water and the scale is inches, indicated values are read directly as "inches of water". If the fluid is mercury, and the scale millimeters, the values are read as "millimeters of mercury". This is an extremely precise measurement if the liquid is of known pure concentration. The properties of pure water and the resistance of earth's gravity are extremely constant values.

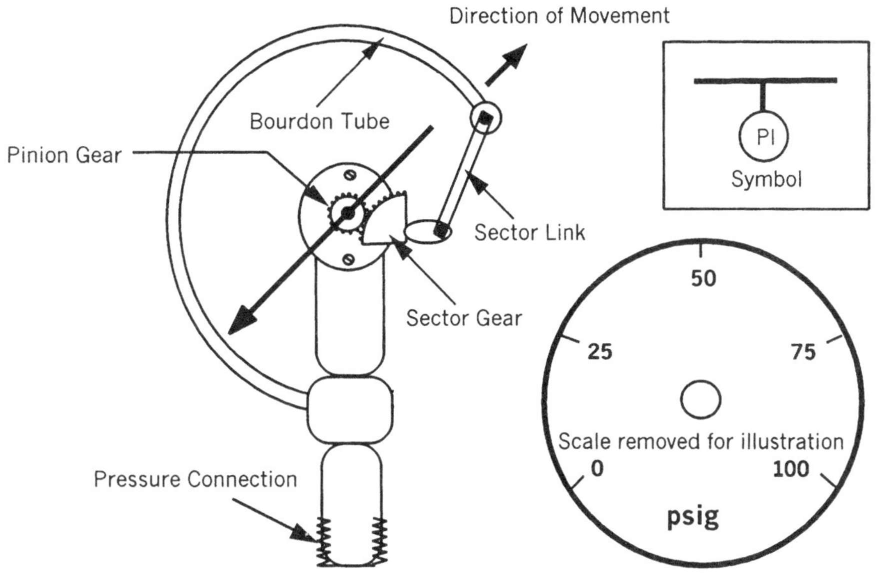

Figure 2.1-4 *Bourbon C-Type Tube Assembly*

Bourdon tube elements, C type, spiral, and helix are elastic deformation elements. They convert fluid pressure into physical displacement. Used in some pneumatic controllers, these elements are most commonly used in local indication pressure gauges. Robert Hooke (1635–1703) developed several principles governing the design of modern pressure tubes. Key principles are stress, strain, and elasticity. Stress is the magnitude of a force applied to an object over the area of an object, strain is the measure of deformation caused by stress, and elasticity is the ability of a material to return to its original shape and size after an applied stress is removed. An object stressed beyond its elastic limit retains deformation after the stress is removed. Bourdon tubes are produced in C-type, spiral, and helix hollow metallic shapes. Bourdon C-type tubes resembles a question mark (?). Spiral tubes are wound flat and helix tubes resemble a coil spring. In each case, as increased pressure overcomes the spring tension of the tube it attempts to straighten or unwind. A metal link transfers this physical motion to a gear assembly. The gear assembly converts the linear motion of the tube into torque (a twisting motion). A pointer affixed to the gear indicates pressure on a calibrated circular scale. Manufactured elasticity and tensile characteristics of the tube determine its sensitivity and operating range. Adjustable linkages, shown in Figure 2.1-4, allow fine-tuning.

Common compression displacement elements include the diaphragm, capsule, and bellows. Much like elastic elements they convert changes in pressure into physical displacement. Elastic elements function as a spring, as pressure changes; they deform and recover within their elastic limit. Compression elements are not designed to deform. Diaphragms act primarily to isolate and transfer changes in pressure from the external process media to an internal instrument fluid, Figure 2.1-6. Diaphragms are constructed as sensitive foils, corrugated metallic disks. They are affixed to the outside of a metal well or cavity. The cavity

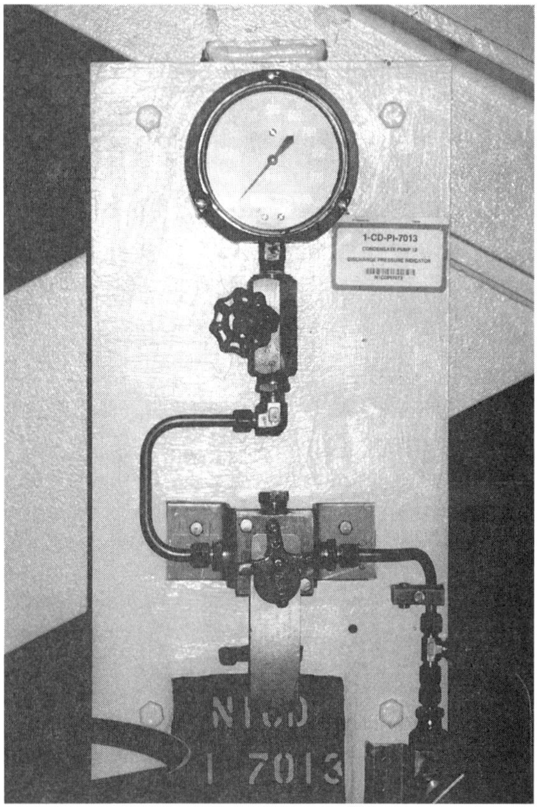

Figure 2.1-5 *Pressure Gauge with Adjustable Pulsation Damper*

is evacuated (all air removed) and filled with instrument oil. Once the cavity is sealed, any compressive displacement of the diaphragm reduces the volume of the cavity. As internal volume of the cavity is reduced, pressure within the cavity increases, and this increase is detected by additional instrumentation connected by capillary tubing to the diaphragm.

Capsules are constructed and operate similarly to diaphragms, however, provide more than isolation and transfer functions. Capsules have an electronic device or mechanical force beam internal to the capsule that converts physical displacement of the foil (compression) into an electronic or physical displacement suitable for monitoring by a transmitter. Figure 2.1-7 shows the arrangement of a differential pressure capsule whose output is the pressure difference between two points.

Bellows are constructed as hollow metallic corrugated cylinders. They convert pressure energy into direct linear motion. Bellows, shown in

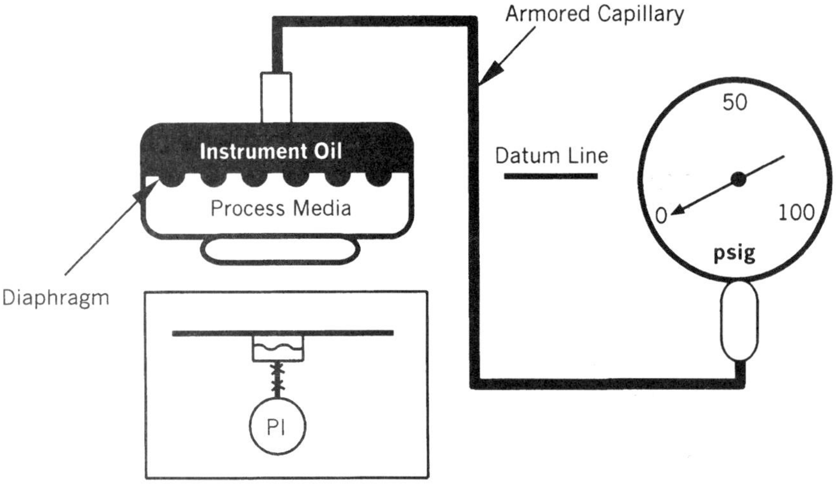

Filled capillary systems are elevation sensitive, if gauge is mounted above or belo diaphragm seal elevation reference (datum line) zero compensation is required.

Figure 2.1-6 *Diaphragm Seal with Capillary Gauge*

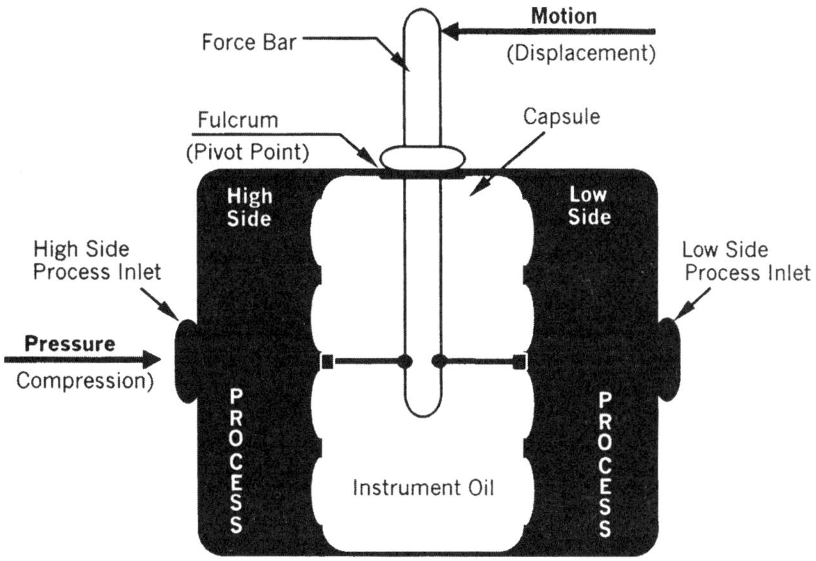

Figure 2.1-7 *Differential Pressure Capsule*

Figure 2.1-8, are typically constructed out of brass, copper, or stainless steel. They are sensitive and fragile and measure only lower pressure ranges, typically limited to less than 30 psig. Some differential pressure gauges contain a set of opposing bellows attached to a torque tube assembly. The difference between the pressures applied to both bellows forces the bellows assembly to deflect. The torque tube converts bellows deflection into a rotational force. A pointer or link arm connected to the torque tube provides differential pressure indication.

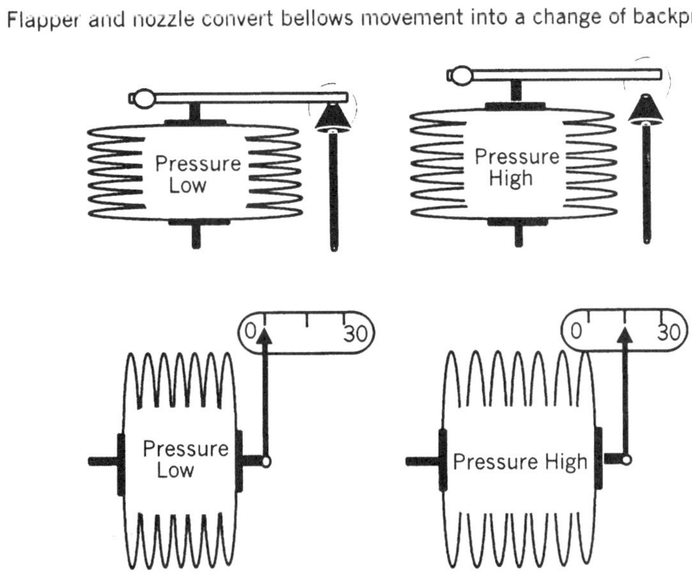

Figure 2.1-8 *Bellows Assembly*

Strain gauges and piezoelectric transducers are electronic pressure elements. Strain gauges convert physical displacement into a change of electrical resistance. The most common application of strain gauges is load cells. They are sometimes used in conjunction with compression displacers to convert motion into resistance. Strain gauges are constructed as a serpentine pattern of wire bonded to a backing. As the wire is stressed the strain causes the wire to elongate. The resistance of a long small diameter wire is greater than the resistance of a short large diameter wire. Therefore, as strain increases, resistance also increases. Piezoelectric transducers convert physical displacement into electric voltage or electric voltage into displacement. The piezoelectric effect is a natural occurrence in many natural and artificial crystals. They are also used in conjunction with other devices.

Pressure sensing elements malfunction for a variety of reasons. The most common failures are over-range, rupture, corrosion/erosion, plugging, loss of fill fluid, and fatigue/brittleness. Figure 2.1-9 illustrates the uncharacteristic bulges or contours of an over-ranged element. Split seams and tears identify ruptures. Melting, pitting, and scarring are evidence of corrosion and erosion. Plugged instruments exhibit a total lack of response to system changes. Loss of fill fluid, Figure 2.1-10, is detected as a lack of response and a hollow or empty sound when the diaphragm is gently tapped. Fatigue and brittleness is detected as a loss of accuracy and lack of repeatability.

Pressure is measured in many different ways. Different units of force and area, as well as different points of reference are used. Typically, the magnitude of the variable measured determines the units and reference point used. To avoid confusion, units are always identified with the scalar value. Pounds per square inch (psi) are used to identify larger magnitudes of measure. Pressures of lesser magnitude are identified as inches of water ("H_2O) or inches of mercury ("Hg). Units for both gauge and absolute pressure are pounds per square inch (psi). The point of reference (0) for gauge pressure (psig) is atmospheric

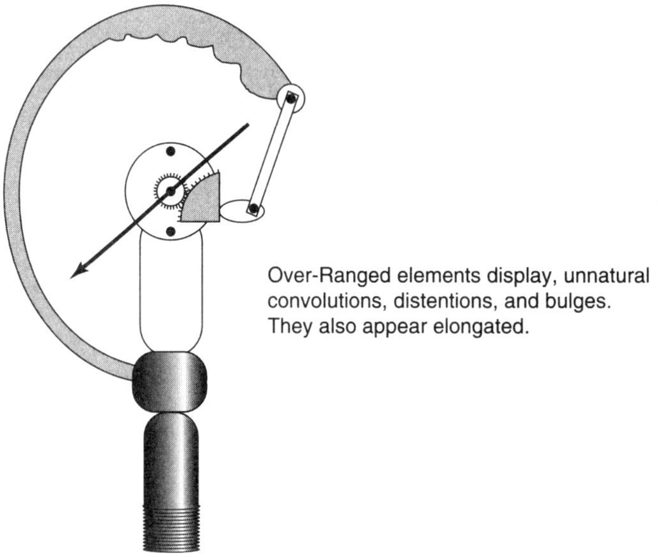

Figure 2.1-9 *Over-Ranged Bourdon Element*

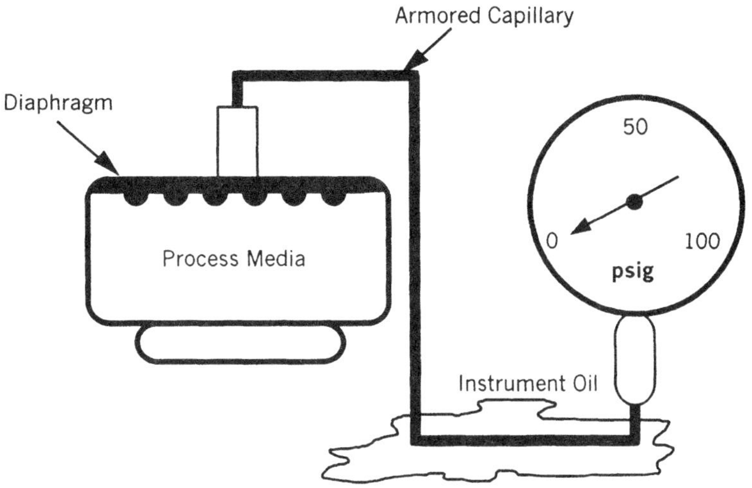

Figure 2.1-10 *Diaphragm Seal with Leak at Capillary*

pressure. Figure 2.1-11, demonstrates the point of reference (0) for absolute pressure (psia) is absolute zero. Pressures less than atmospheric are identified as inches of mercury vacuum ("Hg VAC) or millimeters mercury (mm Hg) absolute.

Conversion Factors

1 pound per square inch (psi) equals:

27.7 inches of water ("H_2O)
2.036 inches of mercury ("Hg)
51.71 millimeters of mercury (mm Hg)

Unit analysis is a method of conversion that assists in the preservation of correct value identities. This procedure incorporates the basic identity law; any number divided by itself is equal to one. The values of conversion are established as statements of ratio equalities. Numeric values are manipulated normally, but the unit identifiers are manipulated as if they were exponents.

Example: to convert 10 feet into inches determine the correct conversion factor:
1 foot is equal to 12 inches therefore: 1 ft/12 inches = 1

State the unknown as a ratio.	10 feet/N	= 1
Establish the ratios as an equality.	1 ft/12 inches	= 10 feet/N
Solve for value N.	N × 1 ft	= 10 feet × 12 inches
	N	= (10 feet × 12 inches)/1feet
Numerically	N	= 120/1 = 120
Solve for Units N.	units N	= feet × inches/feet
Restated	units N	= (feet/1) × (1/feet) × (inches/1)
Reduce units	N	= inches

Combined the solution is 120 inches.

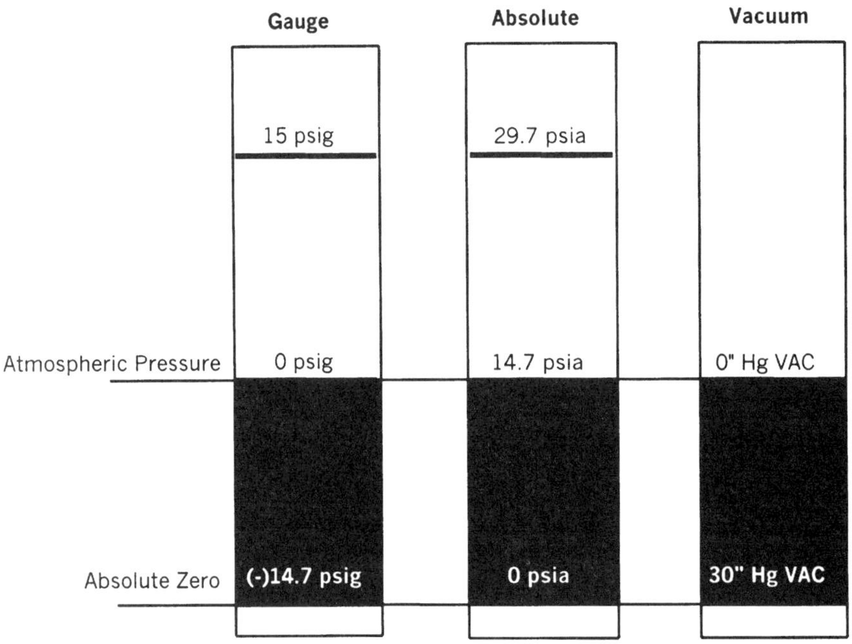

Figure 2.1-11 *Gauge Pressure/Absolute/and Vacuum*

Example: Convert 10 inches of water pressure into millimeters of mercury. 27.7 inches of water is equal to 51.71 millimeter of mercury therefore:

27.7 "H_2O/51.71 mm Hg = 1

State the unknown as a ratio. 10 "H_2O/N = 1

Establish ratios as an equality.

27.7 "H_2O/51.71 mm Hg = 10 "H_2O/N

Solve for value N. N × 27.7 H_2O" = 10 "H_2O × 51.71 mm Hg

N = (10 "H_2O × 51.71 mm Hg)/27.7 "H_2O

Numerically N = 517.1/27.7 = 18.67

Solve for units N units N = "H_2O × mm HG/"H_2O

Restated units N = ("H_2O/1) × (1/"H_2O) × (mm Hg/1)

Reduce units N = mm Hg

Combined the solution is 18.67 mm Hg.

Pressure measuring instruments are often exposed to vibration, pulsation, and corrosion. Pressure snubbers, pigtails, and gauge protectors, as shown in Figure 2.1-12, are often installed between the pressure device and severe processes. Pressure snubbers dampen pressure spikes and pulsations. Setscrews with small orifices drilled through them are often supplied with new gauges. The setscrew is installed into the threaded attachment portion of the bourdon tube. Special snubber devices that resemble pipe-fittings are also used. They are threaded onto the gauge connection between the gauge and process piping. Porous

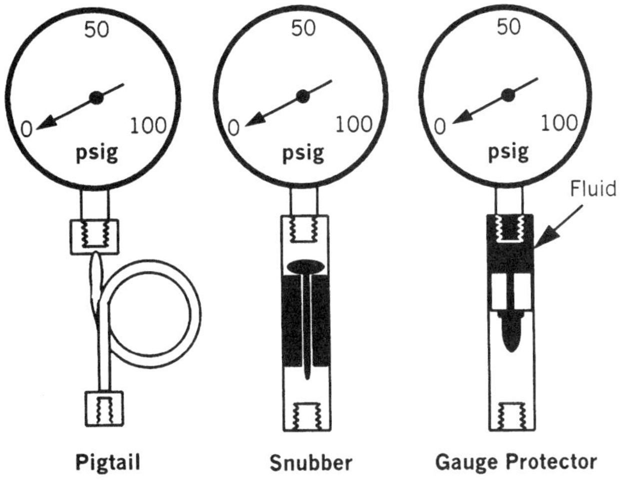

Figure 2.1-12 *Gauge Protectors*

metallic filters or precision piston plungers inside the snubber prevent rapid pressure changes from damaging the gauge. Snubbers effectively dampen damaging pulsation to the gauge, but response time is attenuated, and they are susceptible to plugging. Pigtails are installed between the gauge and process piping to protect the gauge from steam and other harmful vapors. A section of pipe or tubing is formed into a loop. The loop collects and traps condensate to form a seal. The condensate isolates the gauge from extreme temperatures. Gauge protectors are installed on gauges used in extremely corrosive processes. The gauge protector is a diaphragm seal constructed of corrosion resistant material. A special evacuation and fill process is used to connect the gauge to the gauge protector. Disconnecting the gauge from the protector results in a loss of fill fluid and renders the gauge inoperable.

2.2 Temperature

Temperature is the basic measurement of thermal activity, or heat. Heat is the energy that flows between two bodies because of the temperature difference between them. Just as pressure attempts to reach equilibrium within a system, so too does heat. Heat transfers from one object to the next through three distinct processes. These processes, depicted in Figure 2.2-1, are conduction, convection, and radiation.

Conduction requires direct physical contact. Conduction is characterized as a transfer of heat energy throughout a body, without a transfer of substance through the body. Heat energy conducts, as it energizes, a continuous chain of molecules one after another. Conduction continues until heat energy among all adjacent molecules is equalized. An example of conduction occurs as an iron poker is placed among glowing fireplace embers. If the poker remains in the embers for a sufficient amount of time, heat energy at the fire end will transmit through the shaft to the handle.

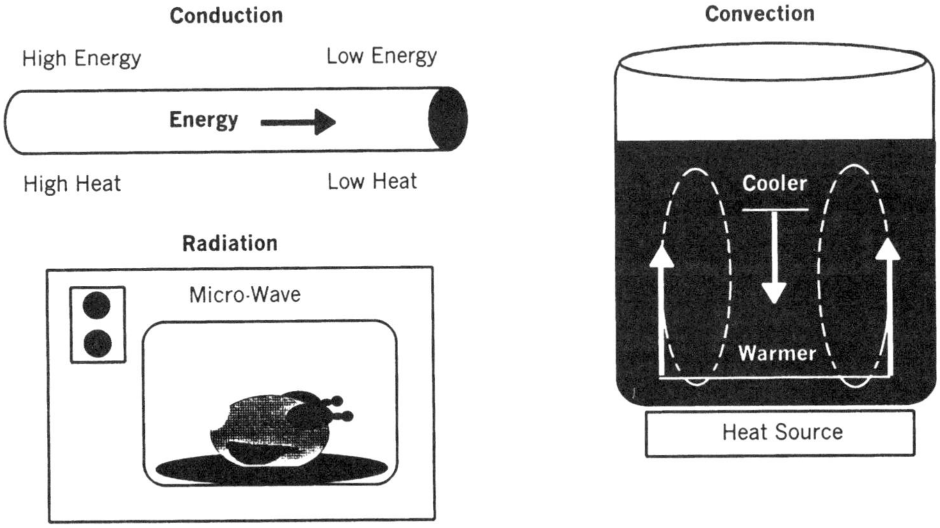

Figure 2.2-1 *Heat Transfer Methods*

Convection occurs as the actual substance containing the heat energy combines, or mixes with another substance. The resulting temperature is again an equilibrium of the combined substances. Convection is a common form of heat transfer among fluids and gasses. Convection transfer occurs when a bit of cold water is added to cool a hot bath. Radiation transfer requires no substance to transmit energy. The heat energy is in the form of electromagnetic waves. The sun warms the earth's surface through radiation transfer.

Two phenomena of heat important to process instrumentation systems are phase changes and thermal expansion. Any pure substance can exist in one of three states; solid, liquid, or vapor. All substances assume their physical state based on specific combinations of heat and pressure. The amount of energy absorbed by a substance during a phase transition or change of physical state is referred to as latent heat. Water in an automobile radiator maintains a liquid state at temperatures above its atmospheric boiling point. The water remains liquid if the radiator cap stays on and keeps the coolant pressurized. If the cap is removed while the temperature is above 212 degrees F, some of the water instantly changes from a liquid state into a vapor and flashes to steam. As a solid substance is heated it eventually reaches its melting point. Additional heat energy can not increase the temperature of a solid above this point. It does however force a phase change of the object from a solid to a liquid. Once the substance becomes liquid it once again increases in temperature in response to heat energy. The temperature continues to rise until the liquid attains its boiling point. At this point the temperature of the liquid does not increase. The liquid does however transition to a vapor.

The second phenomenon, thermal expansion, affects each substance to a different degree. Increasing temperatures generate molecular excitation (increased vibration). These vibrations cause substances to expand and become less dense. Solids become more pliable, liquids

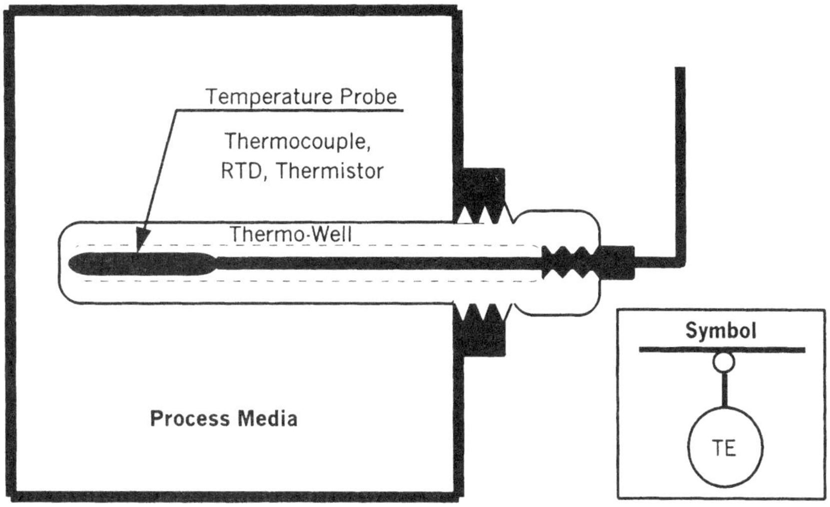

Figure 2.2-2 *Thermo-Well with Probe Installed*

increase in volume, and gasses become lighter. Bi-metallic thermometers and other instruments incorporate thermal expansion as a means of determining temperature. Extended thermal cycling of certain materials induces stress from repeated expansion and contraction.

There are two basic classifications of temperature sensing elements; these are mechanical and electronic. Both physical and electronic sensors many times use thermo-wells. Thermo-wells are heavy-walled metallic rods bored internally to accept the temperature probe, Figure 2.2-2. They provide process isolation, protecting the probe from corrosive and erosive processes and providing the ability to remove and replace probes without breaching the system.

Mechanical temperature elements include the bi-metallic, vapor pressure, and fluid pressure types. Bi-metallic elements are two metals with different thermal coefficients of expansion bonded together. As heat is applied the differences in linear expansion cause the element to distort or bend. This distortion is converted into angular displacement and indicated on a scale by an attached pointer, Figure 2.2-3. The bonded bi-metallic is often wound into a spiral or helix to increase pointer travel and sensitivity. Filled thermal systems, Figure 2.2-4; adhere to the basic equilibrium theories of pressure, volume, and temperature. They convert temperature changes into equivalent pressure changes. Vapor pressure elements work on the phase transition principal. A filled bulb is attached by a capillary to a bourdon, spiral, or helix pressure element. The fluid in the bulb transitions to vapor as heat is applied. This gas or vapor pressurizes the enclosed system and the change in pressure is detected and indicated by the pressure element. Fluid temperature elements are constructed similar to vapor pressure elements. Thcy are filled with fluid that expands at a known rate when heat is applied. The change in internal pressure is detected and indicated by a bourdon, spiral, or helix pressure element.

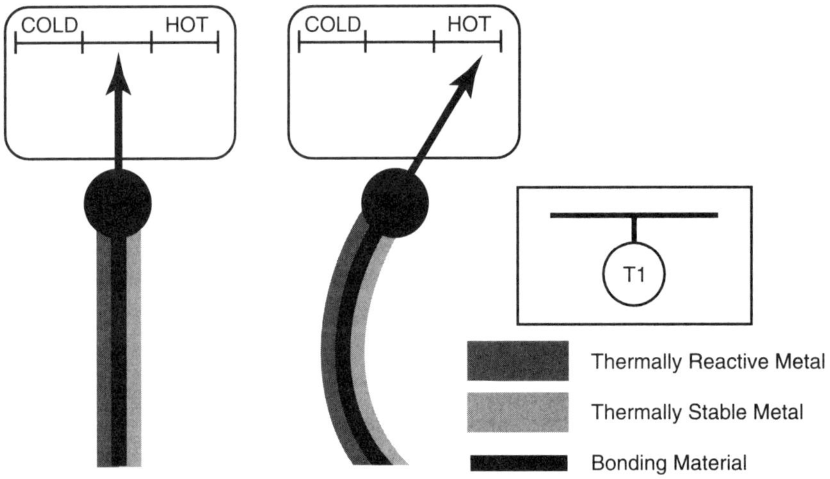

Figure 2.2-3 *Bi-Metallic Temperature Detector*

The primary electronic temperature-sensing elements are the thermocouple, resistance temperature detector (RTD), and thermistor. Pyrometers are used when direct contact is not possible. The thermocouple is constructed of two specially selected dissimilar metals. In 1821, Thomas Seebeck discovered that when two wires of different metal were joined and heated on one end an electric current flowed through the wires. If the wires were disconnected on the unheated junction a voltage was present. Modern thermocouples are used with *electronic bridge circuits to increase flexibility and stability. Thermocouples have a proportional positive temperature coefficient. A temperature increase produces an equivalent voltage increase. Different metallic combinations are used to cover a wide range of temperatures and sensitivities. Figure 2.2-5 lists several thermocouple types and temperature

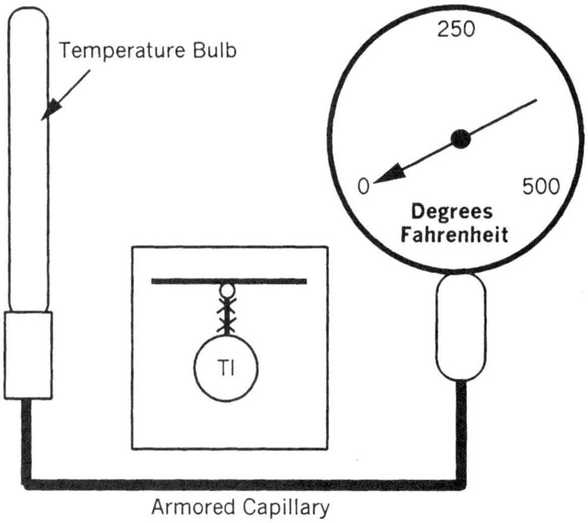

Figure 2.2-4 *Filled Thermal System*

**Electronic bridge circuits detailed in Chapter 3.3*

ANSI Type	Composition		Typical Range °F
	Positive Element	Negative Element	
B	Platinum (20% Rhodium)	Platinum (6% Rhodium)	1600 - 3100
E	Chromel	Constantan	32 - 1600
J	Iron	Constantan	32 - 1400
K	Chromel	Alumel	32 - 2300
R	Platinum (13% Rhodium)	Platinum	32 - 2700
S	Platinum (10% Rhodium)	Platinum	32 - 2700
T	Copper	Constantan	-300 - +700

Figure 2.2-5 *Thermocouple Table*

ranges. Resistance temperature detectors, RTDs, are constructed as a spiral or helix wound wire or a metallic film of pure metal. As the metal is heated the resistance increases. A bridge circuit detects the change of resistance. Like the thermocouple the RTD exhibits a proportionately positive output. Thermistors are thermally responsive resistors. They are man-made semi-conductor devices, and produce a negative exponential output response. Transistor or operational amplifier circuitry typically detects the change in resistance.

Mechanical temperature sensors are prone to zero drift errors over time. Due to the simplicity of design they generally do or do not respond to temperature changes. Malfunction generally dictates replacement. Electronic sensors are fairly reliable. Malfunction of electronic sensors, other than physical damage, is usually a failure of associated electronic circuitry.

Similar to pressure, there are several units of measure used to identify temperature. The common units are degrees Celsius and Fahrenheit. Kelvin, and Rankin are used for absolute measurement. Figure 2.2-6 identifies the three points of reference that assist in the comparison of scales. The ultimate point of comparison is absolute zero, the point totally devoid of all heat energy where molecular motion stops. At this point the Kelvin and the Rankin scale indicate 0°, the Celsius scale -273°, and the Fahrenheit scale -459.7°. The other two points are the transition points of pure water at sea level (one atmosphere). The freezing point of water is 32° Fahrenheit and 0° Celsius. The boiling point of water is 212° Fahrenheit and 100° Celsius. Between the freezing point and boiling point of water a change of 1.8 degrees Fahrenheit is equivalent to a change of 1.0 degree Celsius. Fahrenheit and Rankin indicate the same amount of temperature change per degree. The primary difference is the point of reference; Rankin scale begins at absolute zero. Celsius and Kelvin also indicate the same amount of temperature change per degree. The primary difference is again the point of reference; Kelvin scale begins at absolute zero. The conversions for Fahrenheit and Celsius are, $°F = (1.8)(°C) + 32$ and $°C = (°F - 32)/1.8$

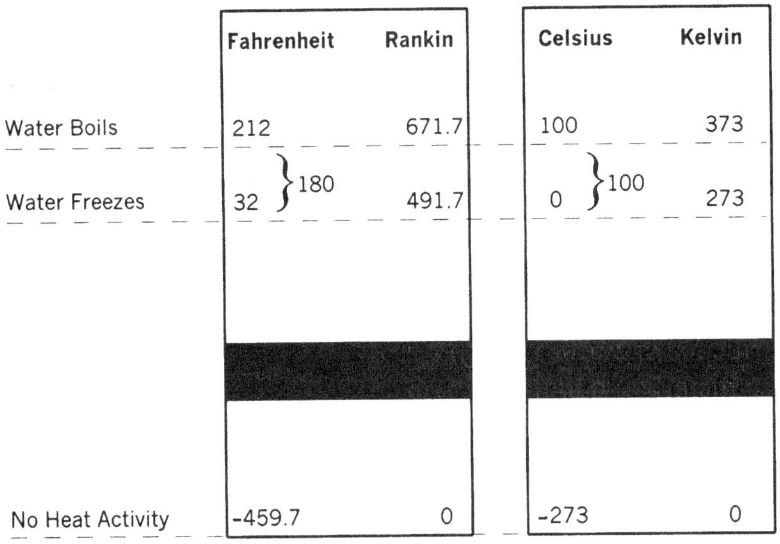

Figure 2.2-6 *Temperature Scale Comparison*

2.3 Level

Pressure, temperature, elevation, and density of the media monitored, all effect level indication. Liquid level, just as pressure and temperature, attempts to maintain equilibrium throughout a system. Density is defined as mass per unit volume. Temperature effects fluid density. Increasing temperatures decrease density as the volume increases but the mass remains the same. Since differences in elevation effect level indications, the point of reference (datum line) is important to consider in level measurement. The lower sensing tap is the reference point for "0" percent level, and the upper sensing tap is the reference point for "100" percent level. Calculations to correct or compensate for elevation differences are called

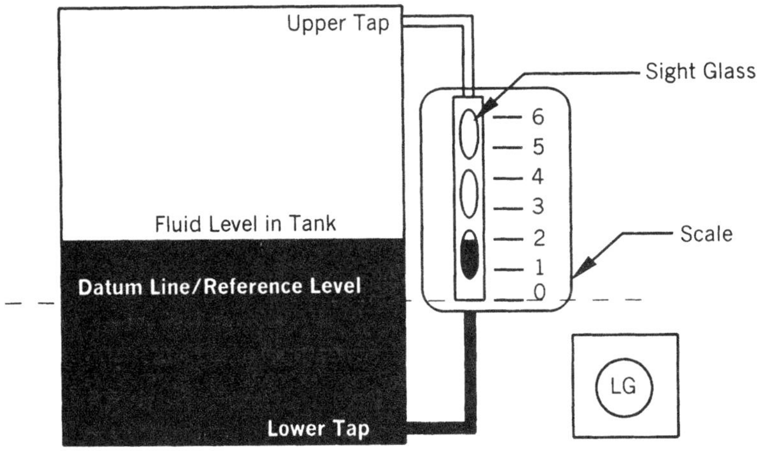

Figure 2.3-1 *Sight Glass Assembly*

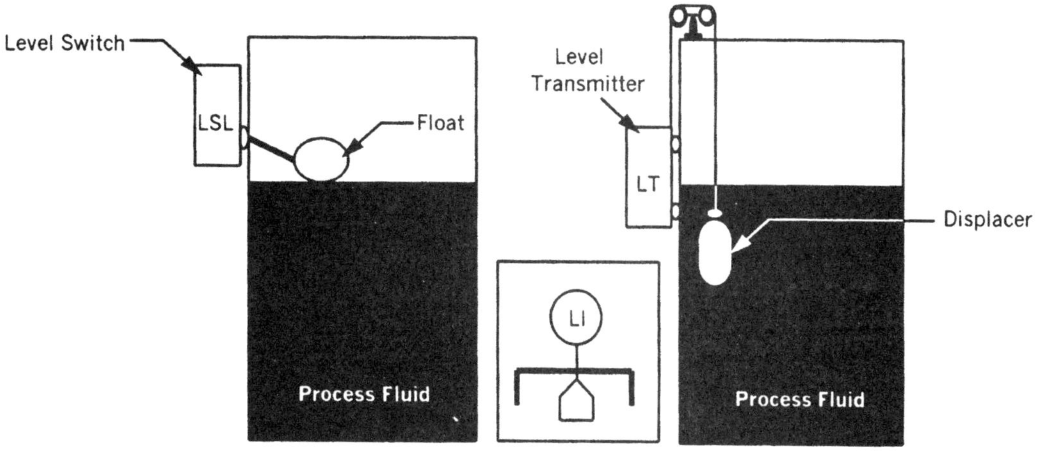

Figure 2.3-2 *Float and Displacer Comparison*

head correction factors. Inches of water and feet of water are the standard measurement units for liquid head.

Common types of level sensing elements include sight glasses, floats, displacers, and differential pressure transmitters. A sight glass is a section of transparent material, usually glass, piped in parallel to a vessel, tank, or column. As the fluid level in the vessel changes, the level in the sight follows. In Figure 2.3-1, the tank level is read directly from a scale etched or attached to the glass tube.

Displacers and floats are sealed balls or cylinders of known volume and density. They both adhere to the Archimedes' principle; the buoyancy of an immersed or partly immersed object is equal to the quantity of fluid it displaces. Floats are designed to remain at the surface of a fluid. Displacers are designed to attain a state of buoyant equilibrium at a known submerged level. Figure 2.3-2 illustrates the comparison of a float to a displacer. Each is attached to a tape, chain, or rod. A change in fluid level causes the float or displacer to rise or fall proportionally. The tape, chain, or rod transfers the motion to support instrumentation that converts motion into an electric, electronic, or pneumatic equivalent signal suitable for transmission. Figure 2.3-3 shows a sump pump and motor with a float control.

Differential pressure transmitters are often used to detect level. Pressure sensitive level instruments are installed with upper and lower sensing taps. Any process pressure in the vessel monitored is applied equally to opposite sides of the instrument and is therefore negated (algebraically summed to zero). Level measurement instrumentation is only sensitive to changes in level between the upper and lower range of the instrument. A tank indicates empty if the fluid level in the tank is below the lower sensing tap, and therefore below the measurement range of the level instrument. As in Figure 2.3-4, zero range suppression compensation is sometimes required to correct the indicated value. Density is the ratio of a fluids mass divided by its volume. Relative density (specific gravity) is the ratio of a fluids

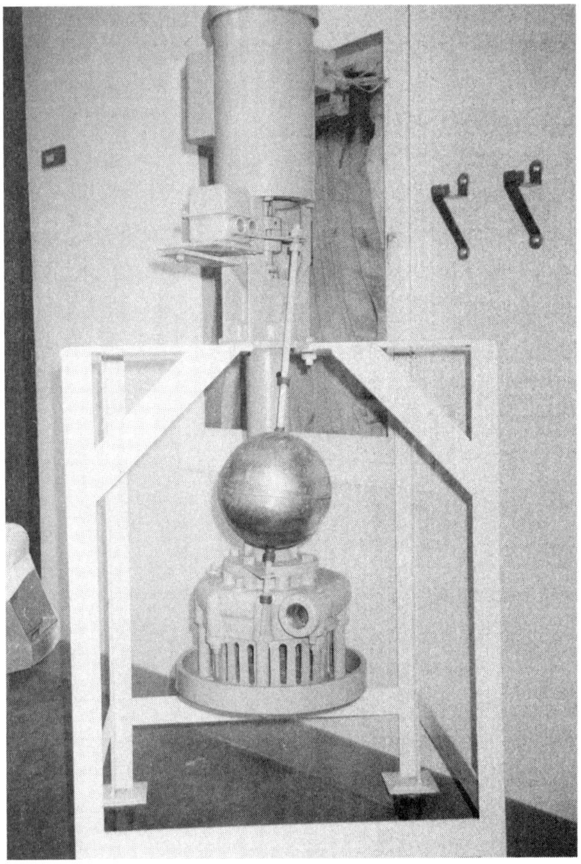

Figure 2.3-3 *Sump Pump with Float Switch*

density divided by the density of water. Pure water has a specific gravity value of one. A correction factor is required for fluids that are more or less dense than water. Several comparisons are provided in Figures 2.3-5, 2.3-6, and 2.3-7. A fluid twice as dense as water has a specific gravity value of 2. A fluid half as dense as water has a specific gravity value of 0.5. The transmitters are piped directly to the vessel with an upper and lower sensing tap. Static pressure is equalized and only changes in level are detected as the algebraic difference between the total high side (positive) value and the total low side (negative) value. Figures 2.3-8, and 2.3-9 clarify this. Upper tap sensing lines are sometimes backfilled with a reference fluid to prevent over range damage, or reverse the operation of the sensor capsule. Specific applications are developed in Figures 2.3-10, 2.3-11, and 2.3-12.

Another level sensing application of differential transmitters is the bubbler tube. In Figure 2.3-13, tubing is mounted vertically inside the tank with the outlet very near the bottom. An air regulator applies a small fixed volume of air to the tube. Air pressure in the tube increases until it is equal to the hydrostatic pressure at the exit point of the tube. Air pressure in the tube remains equal to the hydrostatic backpressure as long as bubbles con-

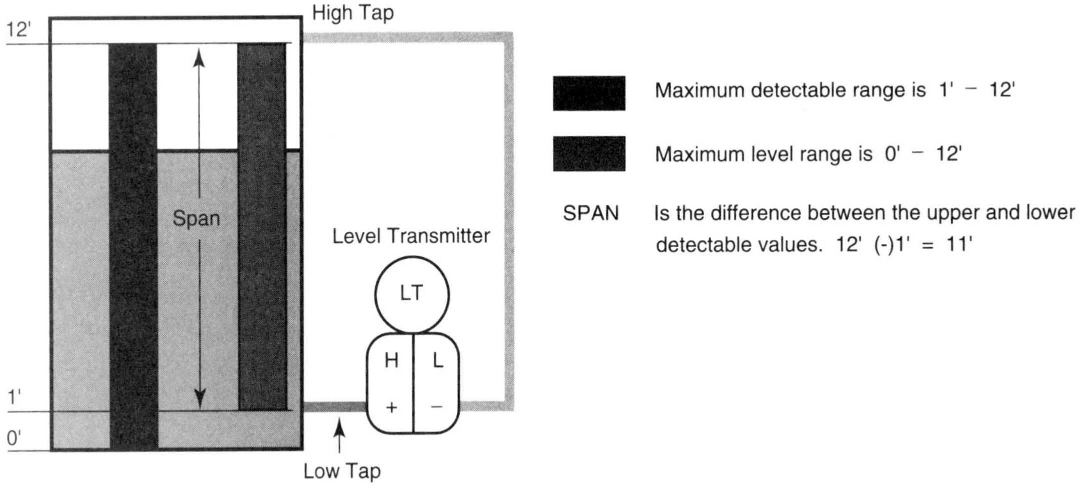

Figure 2.3-4 *Level Transmitter Detectable Range*

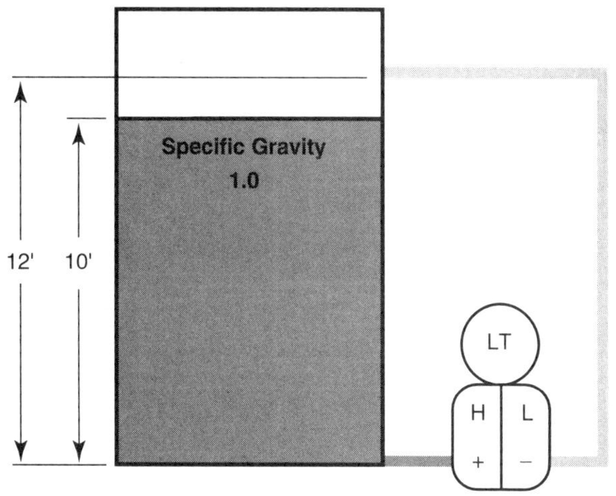

Maximum detectable range is 0 to 12' or = 0 − 144"

Liquid level is 10" x 12" = 120" H_2O

No density correction required if liquid is same density as water (s.g. 1.0)

Span is 144" Measured value is 120"

Figure 2.3-5 *Range and Level Specific Gravity 1.0*

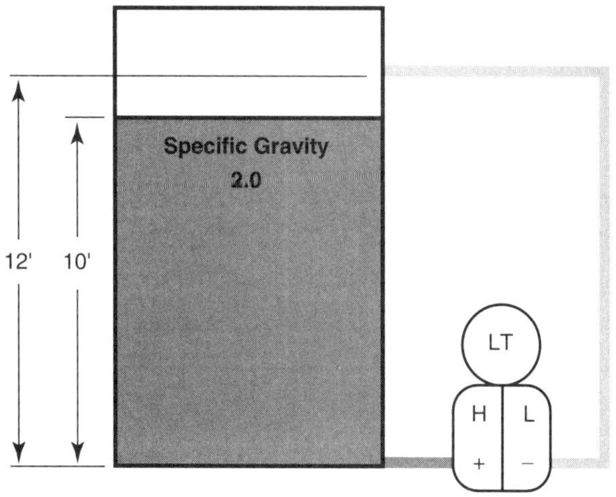

Maximum detectable range is 0 to 12' or = 0 − 144"

Liquid level is 10' x 12" = 120"

Density correction is 120" x 2 = 240" H_2O

Span is 288" Measured value is 240"

Figure 2.3-6 *Range and Level Specific Gravity 2.0*

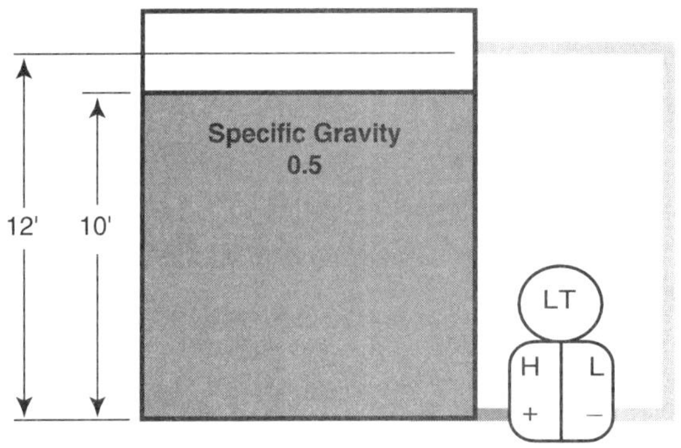

Maximum detectable range is 0 – 144"

Density correction is 144" x 0.5 = 72" H_2O

Liquid level is 10' x 12" = 120"

Density correction is 120" x 0.5 = 60" H_2O

Span is 72" Measured value is 60"

Figure 2.3-7 *Range and Level Specific Gravity 0.5*

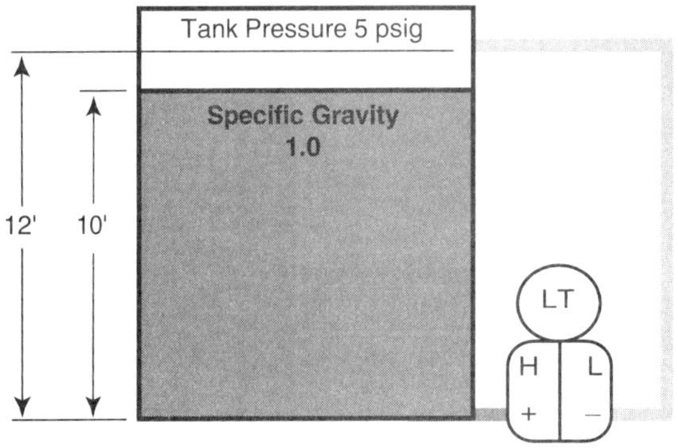

Maximum detectable range is 0 – 144"

Liquid level is 10' x 12" = 120" H_2O

Tan pressure is detected as equal and opposite by transmitter

No correction is required

Span is 144" H_2O Measured value is 120" H_2O

Figure 2.3-8 *Pressurized Vessel Level High and Low Tap*

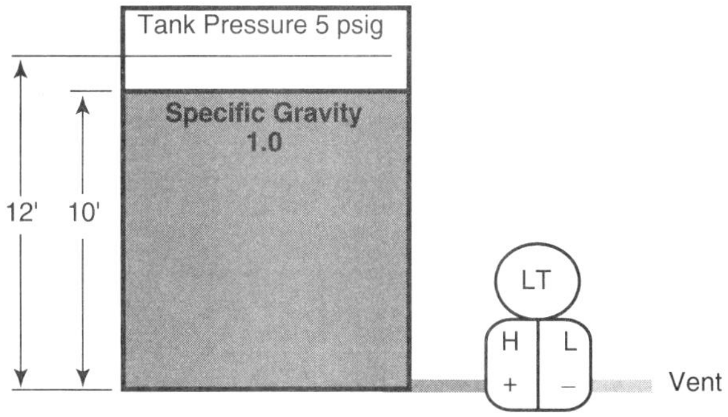

Maximum detectable range is 0 – 144"

Tank pressure is 5 psig x 27.7 = 692.5 H_2O

Corrected range is 692.5 – 836.5" H_2O

Liquid level is 120" + 692.5" = 812.5"

Span is 144" H_2O

Figure 2.3-9 *Pressurized Vessel Level (Low Tap)*

tinue to exit the tube. A transmitter connected to the tube detects the amount of backpressure. Proper level indication is obtained when the air bubbles exit the tube one at a time with stable periodicity (evenly spaced). The flow rate is adjusted correctly when the flow meter indication sphere appears to bounce slightly within the glass cylinder.

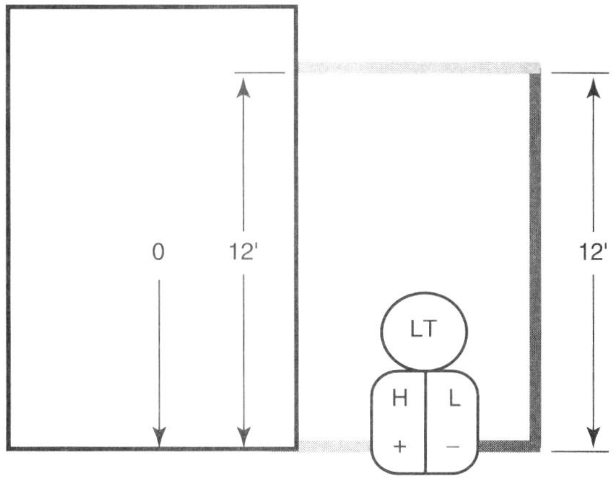

Maximum detectable range is 0 to 12' x 12" = 0 – 144"

Liquid level is 0' x 12" = 0" H_2O

Low side filled reference is always 12' x 12" = 144"

Calibration range is (-)144" to 0" H_2O Span remains 144" H_2O

Measured value is algabraic difference at transmitter 0" – 144" = (-)144" H_2O

Figure 2.3-10 *Vessel Level Low with Filled Reference Leg*

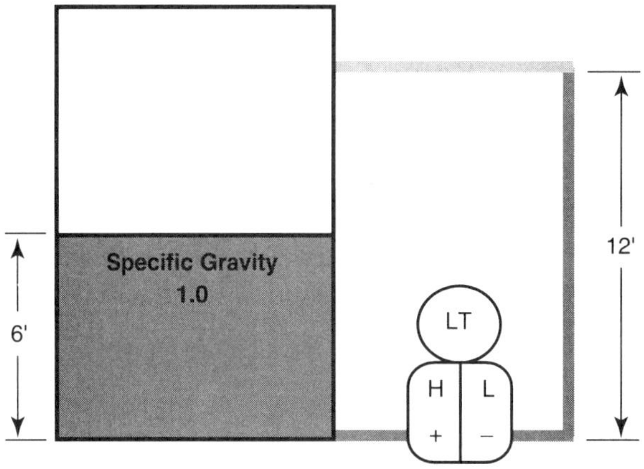

Maximum detectable range is 0 to 12' x 12" = 0 – 144"

Liquid level is 6' x 12" = 72" H_2O

Low side filled reference is always 12' x 12" = 144"

Calibration range is (-)144" to 0" H_2O Span remains 144" H_2O

Measured value is algabraic difference at transmitter 72" – 144" = (-)72" H_2O

Figure 2.3-11 *Vessel Mid-Level with Filled Reference Leg*

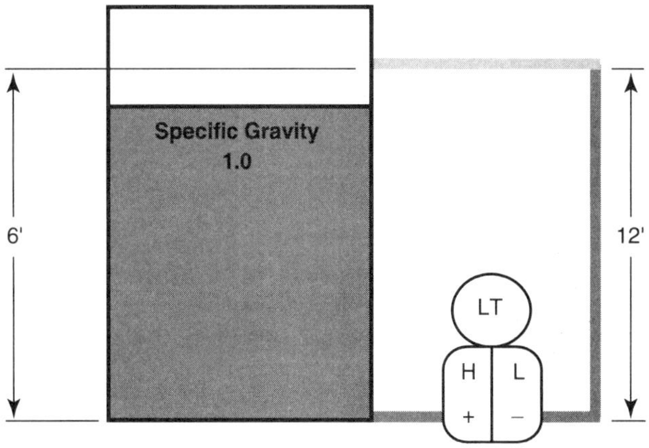

Maximum detectable range is 0 to 12' x 12" = 0 – 144"

Liquid level is 12' x 12" = 144" H_2O

Low side filled reference is always 12' x 12" = 144"

Calibration range is (-)144" to 0" H_2O Span remains 144" H_2O

Measured value is algabraic difference at transmitter 144" – 144" = 0" H_2O

Figure 2.3-12 *Vessel Level Full with Filled Reference Leg*

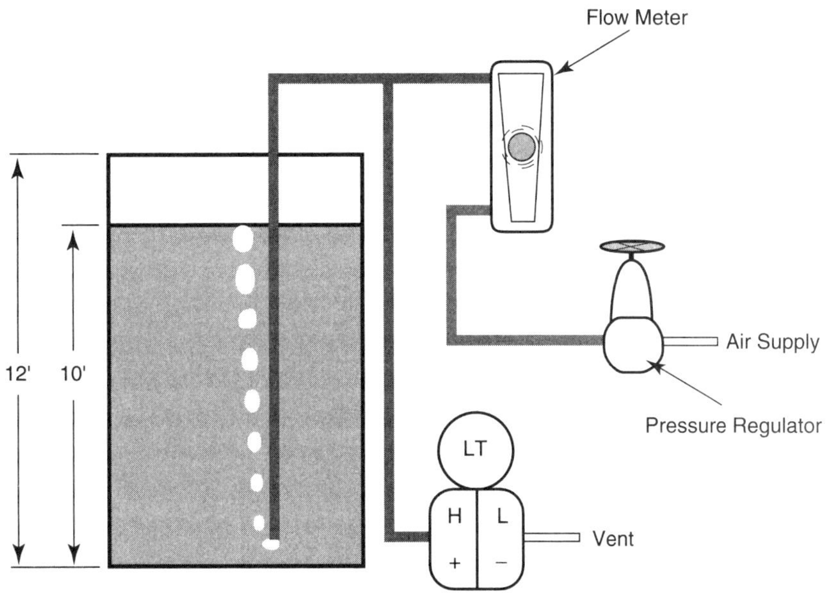

Figure 2.3-13 *Air Bubbler System for Level Detection*

There are several electronic level instruments. The most common are capacitance probes, ultrasonic probes, and radiation transmitters. Capacitance probes are direct contact level sensors. In Figure 2.3-14, two parallel conductive rods extend vertically into the fluid. The rods are fed an excitation voltage. The capacitance between the rods is determined by the amount of charge (excitation), distance between the rods, and the dielectric strength of the material between the rods. Since the charge and distance are fixed, the only variable is dielectric strength. All fluids have much higher dielectric constant values than air. The level of fluid at maximum rod submersion creates maximum dielectric coupling and maximum capacitance. As the level of fluid in the vessel decreases, the dielectrically coupled area between the two rods decreases and capacitance drops.

Ultrasonic and radiation level detectors are non-contacting probes. Ultrasonic transmitters emit a pulsed frequency signal of known wave velocity from the top of the vessel into the perpendicular surface of the fluid, Figure 2.3-15. To determine wave velocities multiply the signal wavelength by the signal frequency. The time delay between the transmission of the original pulse and detection of the reflected signal is proportional to the distance the pulse travels. The elapsed time represents the entire wave cycle, from the transmitter to the fluid and back to the receiver. The actual distance from the transmitter to the fluid is one half of a cycle. Since ultrasonic probes reference level from the top down, they actually detect the amount of fluid not in the vessel. Electronic circuits in the transmit/receive module perform the necessary calculations and conversions. The signal must approach the fluid surface perfectly perpendicular for maximum signal strength.

Radiation transmitters are also non-contacting level detectors. They are two-piece units consisting of a radiation transmitter (source) and a corresponding radiation detector. Radiation

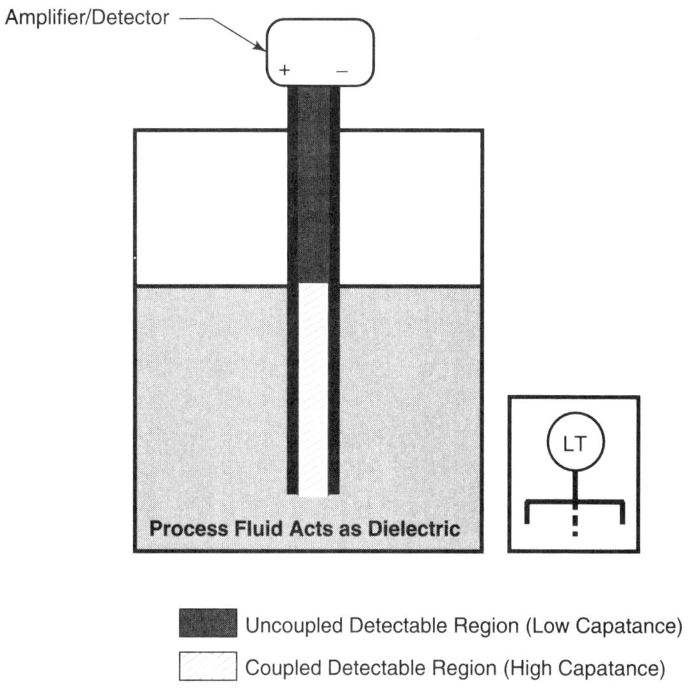

Figure 2.3-14 *Capacitance Level Detector*

travels relatively undiminished in free space. Any fluid present absorbs radiation at a rate specific to the properties of that fluid. The source and detector are placed at the same level directly across the tank from each other. The transmitter emits gamma waves at a fixed known rate. The difference between the energy transmitted and the energy received is representative of the amount of shielding or fluid between the transmitter and detector. High radiation levels indicate an empty vessel. Lower radiation levels indicate the presence of

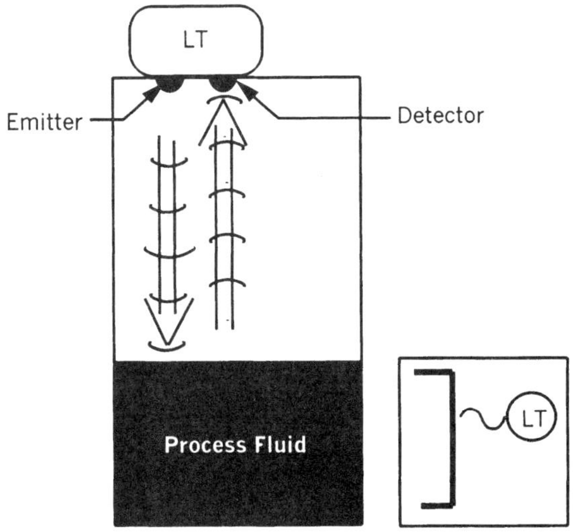

Figure 2.3-15 *Ultrasonic Level Detector*

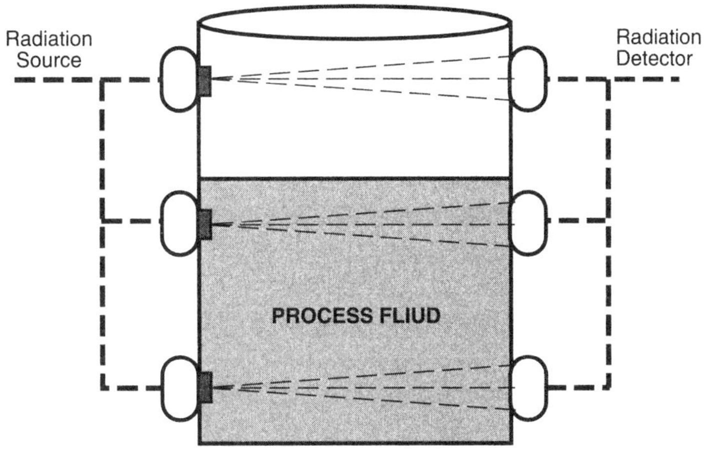

Figure 2.3-16 *Radiation Level Detectors*

fluid between the transmitter and receiver. Since each transmitter/receiver pair only determine the presence or absence of fluid within a narrow range, several pairs are required to track significant level fluctuations, Figure 2.3-16. Although not considered immediately hazardous, low-level radiation sources do require special handling, usage, and disposal considerations.

Sight glass, float, and displacer sensing element malfunctions are usually leaks and sticking floats. Pressure element malfunctions include loss of reference leg, plugged sensing lines, loss of fill fluid and improper bubbler adjustment. Capacitance probes are susceptible to corrosion, galvanic erosion, and do not work with non-conductive fluids. Ultrasonic and radiation element problems are typically detector/transducer failures and misalignment. Ultrasonic signals must approach the fluid surface perfectly perpendicular for maximum signal strength. Radiation transmitter/detector pairs mounted perfectly diametric in the same plane generate maximum signal strength.

2.4 Flow

Flow-rate measurement monitors the instantaneous volume of process media transferring per unit of time. Total flow is the accumulated volume transferred over a period of time. Flow rate terms include both units of volume, (gallons, cubic feet, cubic centimeters), or mass (moles, pounds), and units of time (hours, minutes, seconds). Typical flow rate units are gallons per second and cubic feet per minute. Flow rate measurements of volume require compensations for physical and chemical factors. Pressure, density, viscosity, and velocity, all affect flow measurement. Daniel Bernoulli (1700–1782) developed the principles of flow measurement in use today. Bernoulli's fundamental dynamics of flow state that there is an inverse relationship between pressure and velocity. If fluid velocity increases pressure decreases and if velocity decreases pressure increases.

Additional important concepts pertain to incompressible fluids. The volume, mass, and energy of a fluid in motion maintain continuity. Simply stated, whatever enters one end of a pipe is equivalent to whatever exits the other. The relationship of flow rate, velocity, and differential pressure are simplified by the equation Q = K/DP. The principle statement is a derivation of Bernoulli's equation for ideal flow of a non-compressible fluid (liquid), $Q = K\,Ao\sqrt{2gh}$. Q is the flow rate (volume per unit of time), K is the efficiency factor (a correction factor obtained from charts or tables), A0 is the area of the orifice in square feet, g is the acceleration of gravity (32 feet/second/second), and h is the difference in pressure across the orifice. Compressible fluid (gas) calculations must include static pressure (the amount of compression). The formula for gas flow is $Q = K\sqrt{hp}$. Q is flow in cubic feet per hour, K is a conversion correction factor (obtained from tables), h is the difference in pressure across the orifice, and p is the upstream static pressure. Static pressure is the actual total system pressure. Differential pressure is the difference in pressure between the upstream and downstream tap. Differential pressure increases with flow rates, but not linearly, as shown in the equation the differential pressure increases with the square of the flow rate.

Mechanical flow sensing elements are the rotameter, orifice plate, venturi, flow nozzle, annubar, pitot tube, elbow tap, weir, and flume. The rotameter, orifice plate, venturi, flow nozzle, annubar, and pitot tube actually detect pressure differential resulting from changes in velocity. The physical effect, depicted in Figure 2.4-1, resembles an hourglass and the point of maximum velocity and minimum pressure is called the vena contracta. Each of these detectors create some restriction to flow. The restriction forces a change in velocity and sensing taps located upstream and downstream of the restriction detect the resultant pressure differential.

Rotameters are used for local indication only. They are constructed of a ball or float located inside a tapered tube, as depicted in Figure 2.4-3. The ball restricts the fluid flow. As pressure increases the ball rises in the tapered tube and allows increased flow through the tube. The equivalent flow rate values are read directly off the tube or an attached scale.

Orifice plates are the most common flow-sensing element, Figure 2.4-4. They are constructed as a metallic disk with a precision hole machined in the center. Orifices are installed

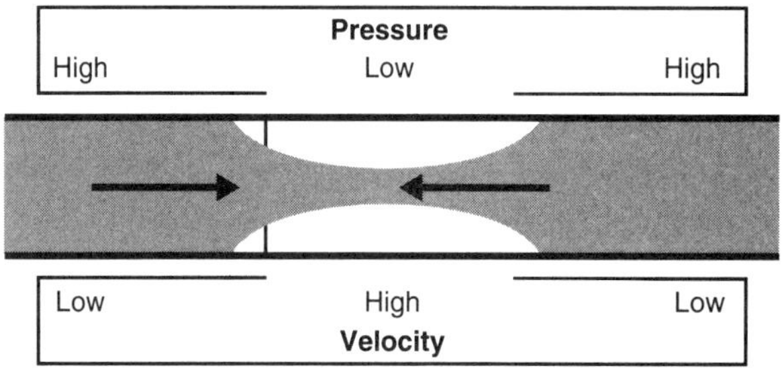

Figure 2.4-1 *Bernoulli Effect*

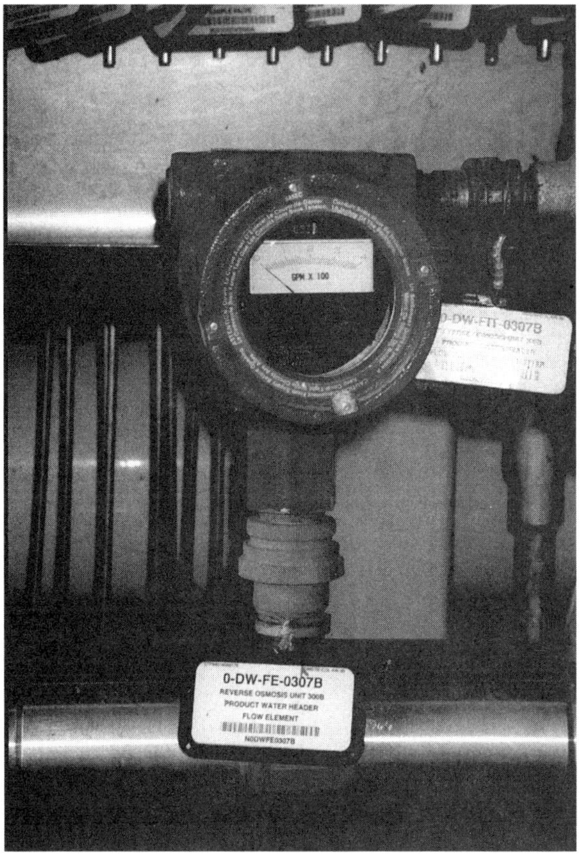

Figure 2.4-2 *Flow Transmitter*

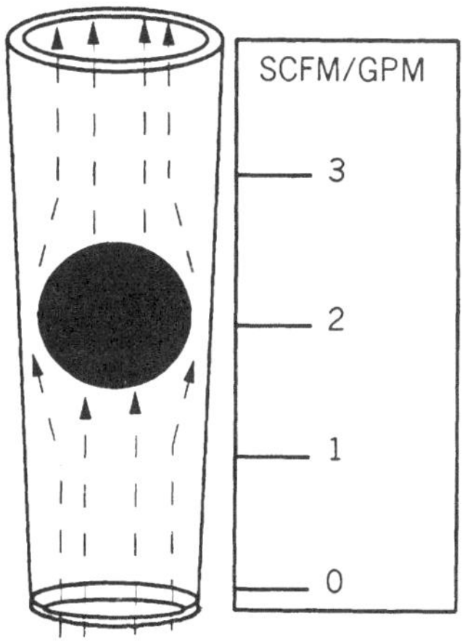

Figure 2.4-3 *Rotameter*

directly into the piping. Interior diameter, exterior diameter, and installation orientation are stamped or engraved on the orifice plate handle. Pressure sensing taps are installed to monitor pressure drop across the orifice, Figure 2.4-5. Weep holes are sometimes provided in the orifice plate to prevent build up in front of the plate.

The venturi and flow nozzle function under the same principles as orifice plates. A venturi is an hourglass shaped restriction in the piping, as demonstrated in Figure 2.4-6. It is less susceptible to plugging than an orifice and creates less restriction to flow. Flow nozzles are basically the upstream section of a venturi.

Annubars and pitot tubes, shown in Figure 2.4-7, are specially shaped tubes with precision-machined sensing vents. The tubes are inserted directly into the flow media perpendicular to the flow. Pressure sensing lines are attached directly to the tubes. As with all flow detection devices the high-pressure area is upstream of the restriction.

Elbow taps are mounted at the inside and outside of ninety-degree bends in the piping, and do not restrict flow. Centrifugal energy of the fluid in motion generates a high-pressure region on the outside of the bend. The difference in pressure is calculated into flow rate. Weirs and flumes are used to measure flow in open channels. Figure 2.4-8 depicts a typical weir. A weir is constructed as a box placed directly in the channel. The upstream area of the box serves to calm the fluid. The actual restriction is a precisely measured weir plate. The theory of measurement derives the cubic volume of fluid flow by multiplying the flow velocity in feet per minute by the square feet area of the water head at the weir notch.

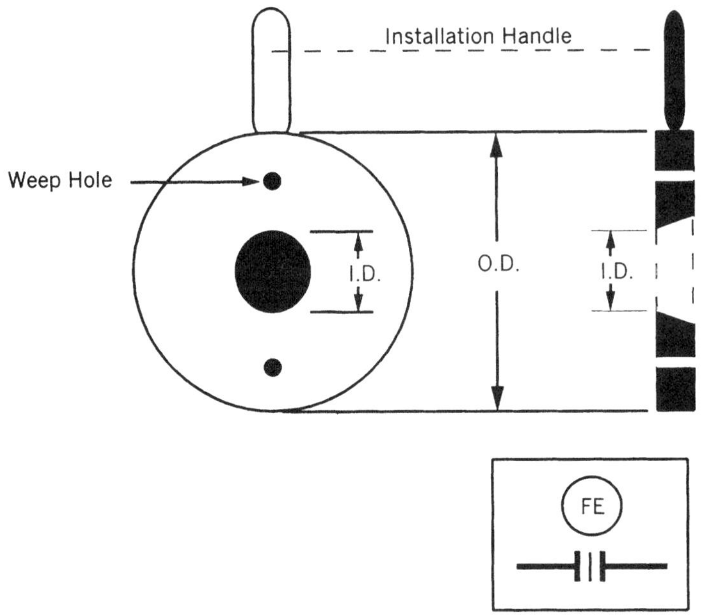

Figure 2.4-4 *Sharp Edge Orifice Plate*

Actual flow figures are usually obtained from a special chart called a nomograph. In 1926 Ralph Parshall documented the use of the flume to determine the flow rate of agricultural water. Flumes are generally constructed of concrete. Figure 2.4-9 depicts a typical flume. The sides of the flume resemble a venturi in shape. The fluid flow is directed through a narrowed section of the channel, since the fluid is not enclosed in piping the restriction results in an increased level in the channel or gate. The change in level is read directly from scaled measuring rods in the fluid, detected by a float or displacer in attached still wells, or sensed as hydrostatic pressure by a pressure device.

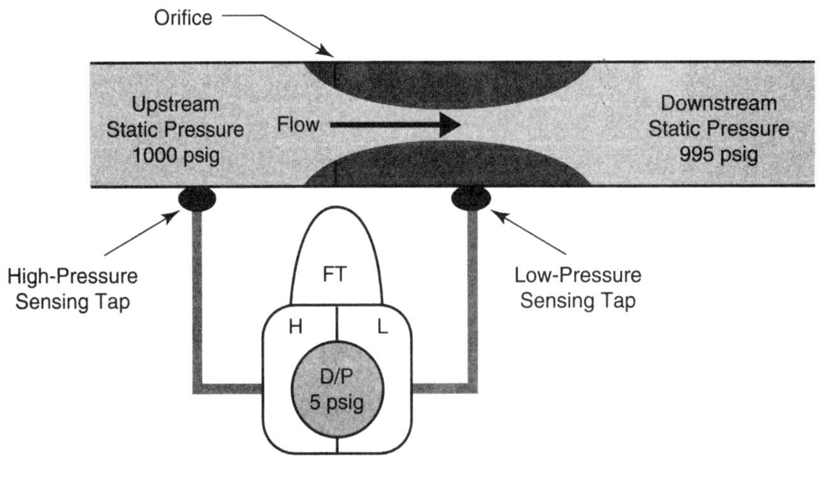

Figure 2.4-5 *Static Pressure and D/P Pressure*

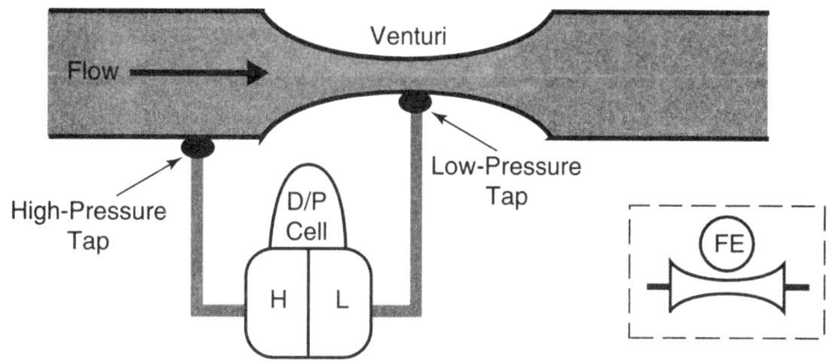

Figure 2.4-6 *Venturi*

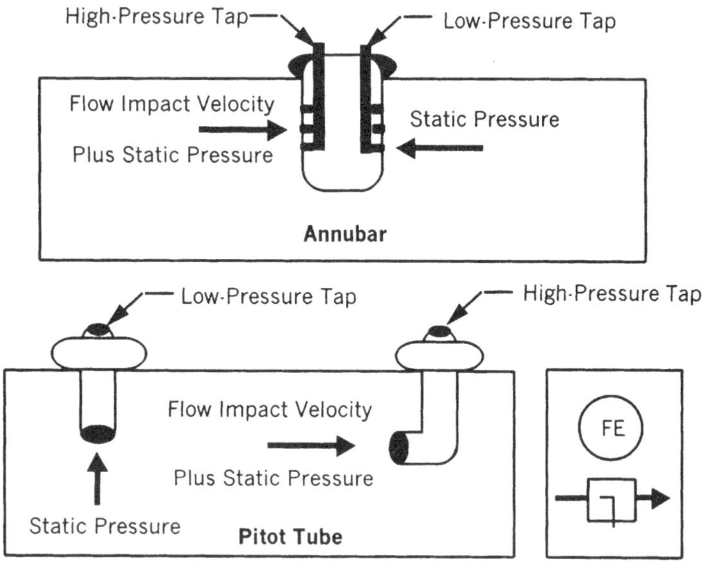

Figure 2.4-7 *Pitot/Annubar*

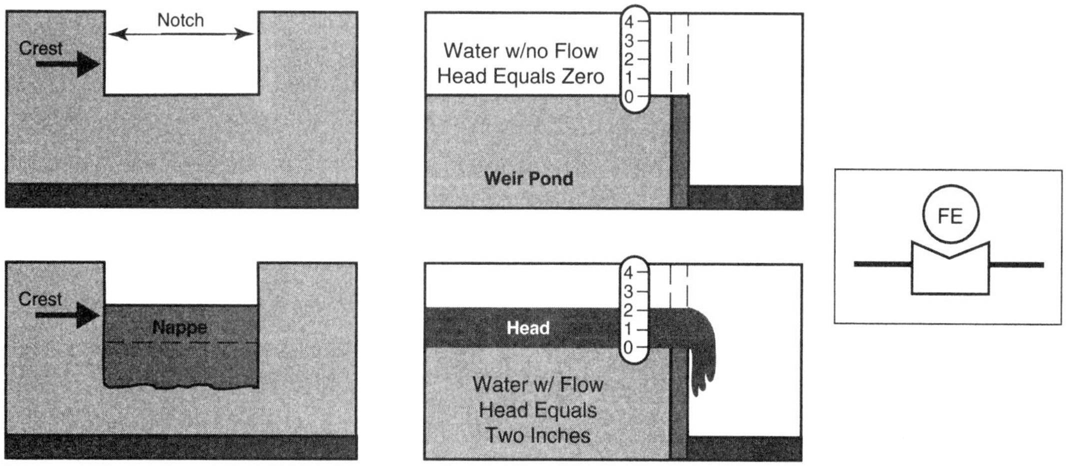

Figure 2.4-8 *Wier*

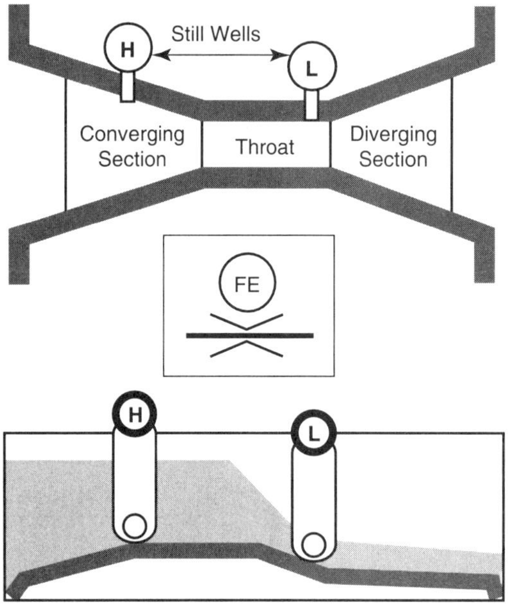

Figure 2.4-9 *Flume*

Electronic instruments also measure flow. The ultrasonic transducer, turbine flow meter, and electromagnet flow meter detect flow velocity. The heated junction converts heat loss into mass flow. The ultrasonic transducer consists of a transmitter and a receiver. The transducers are located on opposite sides of the pipe, linearly offset to transmit diagonally across the flow. Process media assist ultrasonic transmission in the downstream direction and oppose transmission in the upstream direction, Figure 2.4-10. This aiding and abating is detected as a difference in propagation times. Increased flow decreases downstream delay times and increase upstream delay time.

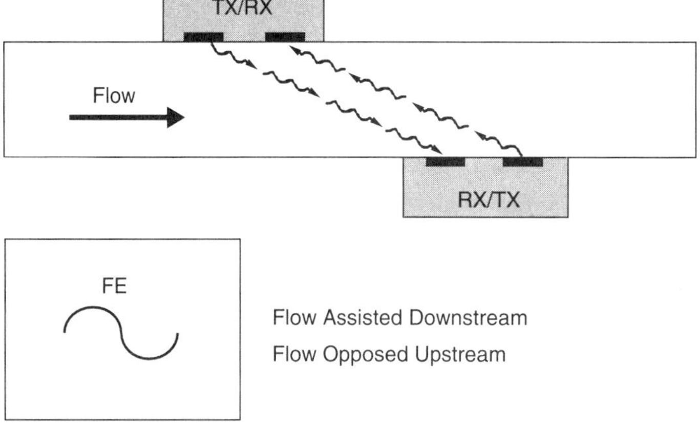

Figure 2.4-10 *Ultrasonic Flow Detector*

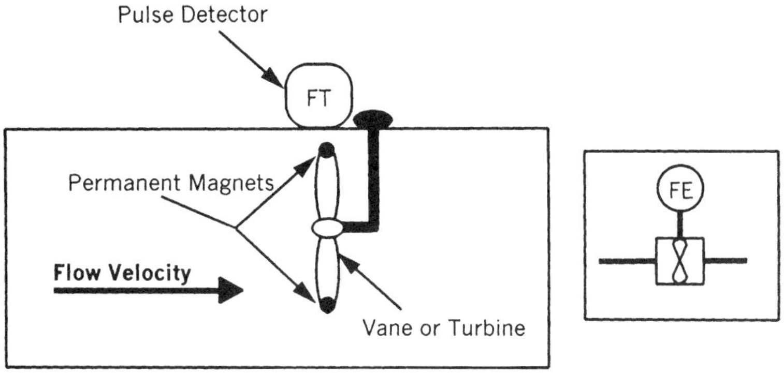

Figure 2.4-11 *Pulse Generator*

The turbine meter shown in Figure 2.4-11, consist of a turbine or propeller mounted inside the pipe facing the flow. Fluid flow departs energy to the turbine causing the blades to spin at a rate directly proportional to the fluid velocity. Magnets attached to the blades produce a pulse as they spin past a detector. The pulse frequency is electronically converted into flow velocity.

Heated junction flow detectors are molar specific. Since they detect the actual mass of flow, they automatically compensate for other variables such as pressure, density, and viscosity. The heated junction consist of an element or wire directly exposed to the process media. Electronic circuits drive current through the wire to hold it at a fixed temperature. Figure 2.4-12 demonstrates the flowing media stripping heat energy from the wire. The amount of current required to offset the heat loss is equivalent to the flow mass absorbing the heat. The changes in heater current are detected and processed by additional circuitry.

Plugging, corrosion/erosion, loss of reference or compensation, and improper installation are the most common flow sensing element malfunctions. Ultrasonic element problems are typically detector/transducer failures and misalignment.

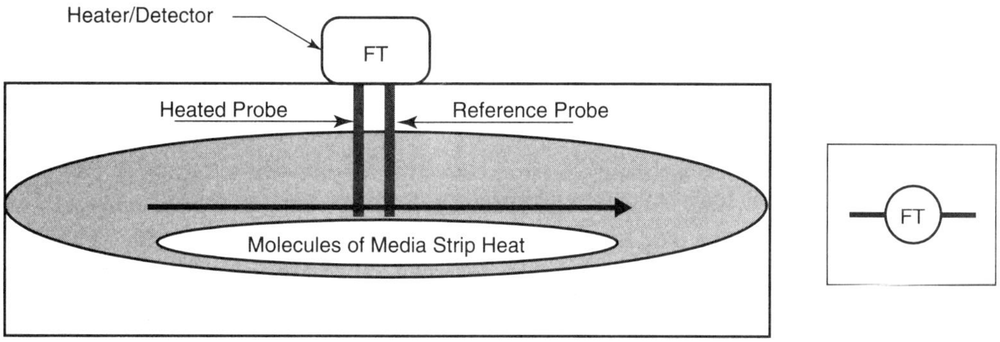

Figure 2.4-12 *Mass Flow Meter*

Chapter Two Summary:

- Primary sensing elements detect and convert energy from the process into a value suitable for measurement.
- Pressure is defined as force exerted over a specific area.
- A change of pressure exerted at any point in a body of fluid at rest is transmitted undiminished to every point in the fluid. Pressure in an enclosed system is dependent on temperature and volume.
- The amount of force exerted by a column of fluid is called hydrostatic pressure.
- The bourdon tube, spiral tube, and helix tube are elastic deformation elements.
- The common compression displacement elements include the diaphragm, capsule, and bellows.
- The strain gauge and piezoelectric transducer are electronic pressure elements.
- The point of reference for gauge pressure (psig) is atmospheric pressure. The point of reference for absolute pressure (psia) is absolute zero. Pressure less than atmospheric is also identified as inches of mercury vacuum ("Hg VAC.).
- Temperature is the basic measurement of thermal activity, heat.
- Conduction is characterized as a transfer of heat energy throughout a body, without a transfer of substance through the body.
- Convection occurs as the actual substance containing the heat energy combines, or mixes with another substance.
- Radiation is in the form of electromagnetic waves; transfer requires no substance to transmit energy.
- Thermo-wells provide process isolation, protecting the probe and providing the ability to replace probes without breaching the system.
- Mechanical temperature elements include the bi-metallic, vapor pressure, and fluid pressure types.
- Vapor pressure temperature elements work on the phase transition principal.
- Fluid temperature elements are filled with fluid that expands at a known rate when heat is applied.
- Three primary electronic temperature-sensing elements are the thermocouple, RTD, and thermistor.
- RTDs and thermocouples have a positive temperature coefficient. Thermistors produce a negative exponential output response.
- Temperature units are degrees Celsius, Fahrenheit, Kelvin, and Rankin. Absolute zero is the point totally devoid of all heat energy where molecular motion stops.
- Pressure, temperature, elevation, and density of the media monitored all affect level indication.
- Liquid level attempts to maintain equilibrium throughout a system.
- Common types of level sensing elements include sight glasses, floats, displacers, and differential pressure transmitters.
- Differential pressure transmitters are installed with upper and lower taps to detect level.

- A correction factor, specific gravity, is required for some level instruments measuring fluids that are more or less dense than water. Pure water has a specific gravity value of one (1).
- Common electronic level instruments are capacitance probes, ultrasonic probes, and radiation transmitters. Capacitance probes are direct contact level sensors. Ultrasonic and radiation level detectors are non-contacting sensors.
- Flow rate measurement monitors the instantaneous volume of process media transferring per unit of time. Total flow is the accumulated volume transferred over a period of time.
- Pressure, density, viscosity and velocity all affect flow measurement.
- There is an inverse relationship between pressure and velocity.
- The point of maximum velocity and minimum pressure is called the vena contracta.
- Static pressure is actual total system pressure. Differential pressure is the difference in pressure across a restriction.
- Mechanical flow sensing elements are the; rotameter, orifice plate, venturi, flow nozzle, annubar, pitot tube, elbow tap, weir, and flume.
- Weirs and flumes are used to measure flow in open channels.
- Electronic instruments used to detect flow velocity are the ultrasonic transducer, turbine flow meter, and electromagnet flow meter.
- Heated junction flow detectors are molar specific and detect the actual mass of flow, automatically compensating for other variables such as pressure, density, and viscosity.

Chapter Two Review Questions:

1. List and identify the primary sensing elements used to detect pressure.
2. List and identify the primary sensing elements used to detect temperature.
3. List and identify the primary sensing elements used to detect level.
4. List and identify the primary sensing elements used to detect flow.
5. Identify primary sensing element malfunctions.
6. Define the term primary sensing element.
7. Define the term pressure.
8. Define the term temperature.
9. Define the term hydrostatic pressure.
10. Define the term static pressure.
11. Define the term differential pressure.
12. Define the term vena contracta.
13. Define the term head correction.
14. Define the terms density and specific gravity.
15. Define the term conduction.
16. Define the term convection.
17. Define the term radiation.
18. Define the term absolute zero.
19. Differentiate between flow rate and total flow.
20. Differentiate between mass and volume measurement as it pertains to flow.
21. Convert 10 psig into psia.
22. Convert 10 psia into "Hg VAC
23. Convert 12 feet H_2O into psig.
24. Convert 277 inches H_2O into psig.
25. Convert 100 degrees Celsius into degrees Fahrenheit.
26. Convert 273 degrees Kelvin into degrees Celsius.
27. Convert 32 degrees Fahrenheit into degrees Celsius.
28. Convert 456 degrees Rankin into degrees Fahrenheit.
29. Calculate level sensor span and range in "H_2O for a D/P transmitter attached to a vessel of pure water with the lower tap installed three feet from the tank bottom and the upper tap installed fifteen feet from the tank bottom.
30. Perform the calculation for the previous system (29) assuming the process fluid is twice as dense as water.

Chapter 3

Transducers and Transmitters

OBJECTIVES

Upon completion of Chapter Three the student will be able to:

- *Define transducers and transmitters.*
- *Describe the operation of a pneumatic transmitter.*
- *State the purpose of the input sensor.*
- *State the purpose of the force balance mechanism.*
- *State the purpose of the pneumatic output relay.*
- *Describe the operation of an electronic transmitter.*
- *State the purpose of the input sensor.*
- *State the purpose of high impedance inputs.*
- *State the purpose of the error detector.*
- *State the purpose of the amplifier/driver.*
- *State the purpose of isolated outputs.*
- *State the purpose of physical displacement transducers.*
- *Describe the operation of a solenoid.*
- *Describe the operation of an electrical switching relay.*
- *State the operation of an LVDT (inches/VDC).*
- *List the major functional components of an LVDT.*
- *State the operation of an eddy current proximeter (mil gap/VDC).*
- *State the operation of a seismic detector (G/volts).*
- *State the operation of a velocity detector (Acceleration/volts).*
- *State Ohms law of equilibrium.*

- *State Kirchoff's law of conservation.*
- *Calculate the combined resistance of a series circuit.*
- *Calculate the combined resistance of a parallel circuit.*
- *Calculate the combined resistance of a series/parallel circuit.*
- *Develop a diagram of a simple bridge circuit.*
- *List two methods of freeze protecting sense lines.*

3.0 Introduction

A transducer is a device that converts one form of signal data into another proportionally equivalent form of signal data. A transmitter is a device that conditions and conveys signal data. These terms are practically synonymous as most modern transducers function as transmitters and most modern transmitters contain integral transducers. Transmitters and transducers are identified by the characteristics of the signal they transmit. The function of the transmitter is dependent on the installation and particular primary element monitored. A pneumatic transmitter with high and low vessel taps functions as a level transmitter. The same pneumatic transmitter with upstream and downstream orifice taps functions as a flow transmitter. In either case the level or flow value detected is transmitted as a proportional and equivalent pneumatic signal. If an electronic transmitter is installed to detect level or flow, the same level or flow value is detected. The transmission data however, is a proportional and equivalent electronic signal.

3.1 Transmitters

Pneumatic transmitters detect and convert a change in pressure or level into equivalent pneumatic signals and retransmit the pneumatic signal data. The most common pneumatic signal is 3–15 psig. Pneumatic transmitters usually incorporate a capsule assembly as the input sensor. The capsule converts any change in input pressure into physical displacement or motion. Capsule displacement is transferred to a force bar. The bar typically is attached to the capsule at one end and a balance beam at the other. The bar levers on a pressure-isolating fulcrum. A flapper nozzle assembly on the balance beam detects displacement of the bar as an error signal, see Figure 3.1-1.

The transmitter is supplied with 20 psig of continuous instrument air (supply). The measured variable value from the primary element is applied to the input sensor (input). The input sensor transfers an equivalent amount of energy to a force balance mechanism. The amount of deviation at the force balance beam is detected by a flapper nozzle assembly, see Figures 3.1-2 and 3.1-3.

The flapper nozzle assembly backpressure is amplified by a pneumatic relay and retransmitted as a pneumatic output signal. Figures 3.1-4, 3.1-5, and 3.1-6 depict several opera-

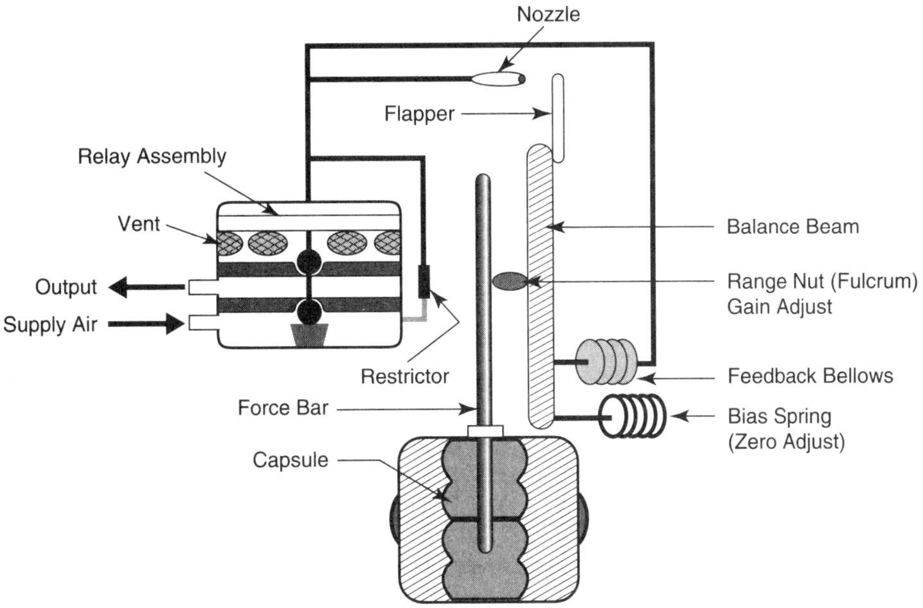

Figure 3.1-1 *Pneumatic D/P Transmitter Diagram*

tional conditions. Supply air flows through a restriction screw and is supplied in parallel to the nozzle, feed-back bellows, and upper chamber of the output relay. The flapper acts to block the nozzle and build backpressure (error signal). The error signal is simultaneously sensed by the feed-back bellows and output relay. The feed-back bellows expands to restore balance to the beam. The relay amplifies the error signal and outputs a representative value of supply air.

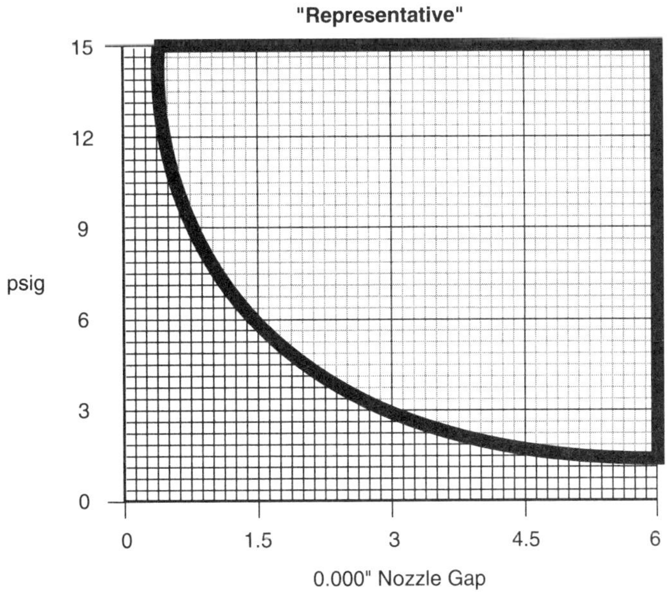

Figure 3.1-2 *Flapper/Nozzle Gap to Pressure*

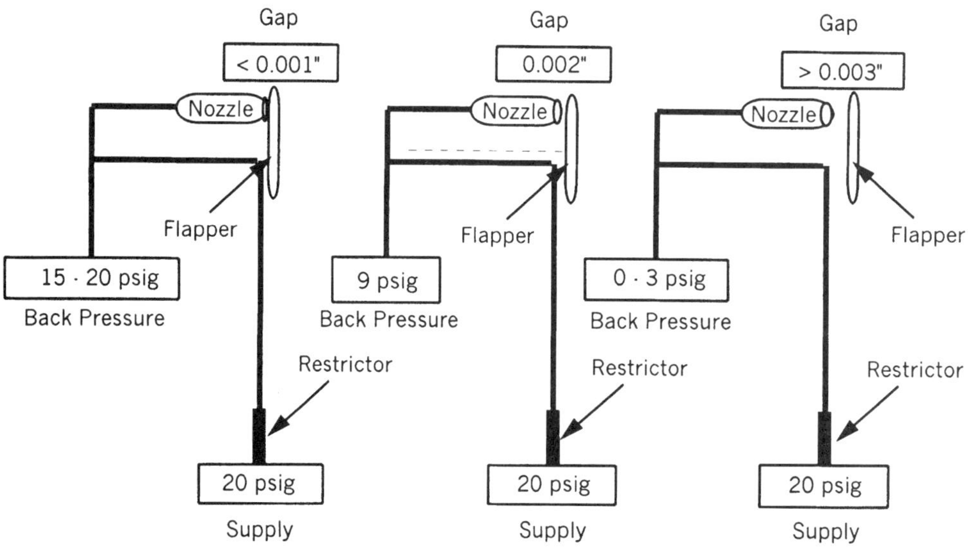

Figure 3.1-3 *Flapper/Nozzle Diagram*

Flexible metallic strips, (flexures), mounted parallel to the force bar suspend the balance beam. The flapper/nozzle is at one end of the balance beam and a feed-back bellows and bias adjustment spring is at the other. Bias spring tension determines the initial balance condition of the balance beam. The beam pivots on an adjustable range nut. Since range nut position determines the fulcrum point, shifting the axial position of the range nut relative to the balance beam determines the mechanical gain or multiplication of force bar motion to balance beam motion, Figure 3.1-7.

Electronic transmitters perform the same basic functions as pneumatic transmitters. The significant difference is the signal output. Pneumatic transmitters output an air signal of

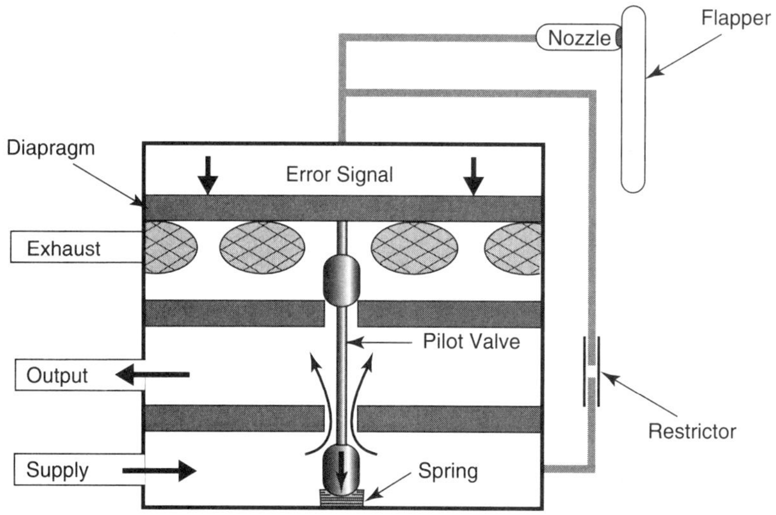

Figure 3.1-4 *Relay Output with Small Nozzle Gap*

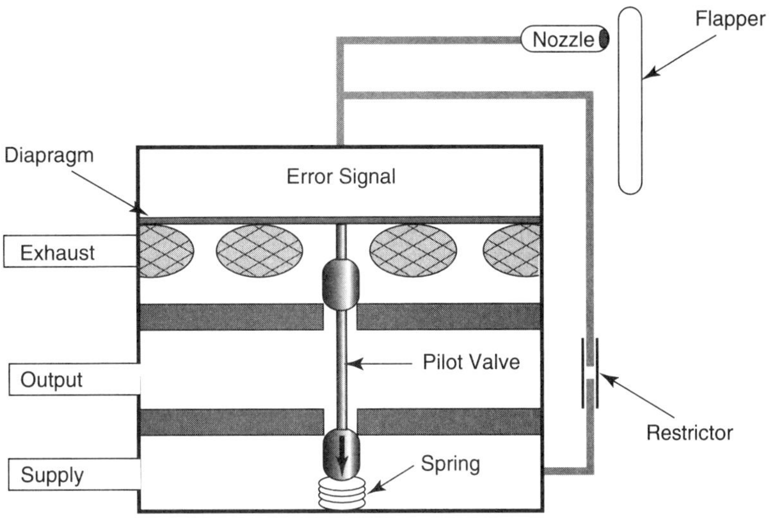

Figure 3.1-5 *Relay with Normal Nozzle Gap*

three to fifteen psig. Electronic transmitters produce an electrical output signal of four to twenty milliamps DC. Intelligent transmitters produce an electrical output of four to twenty milliamps DC with a frequency shift key data transmission signal superimposed. The frequency signal oscillates equally above and below the analog signal and has no effect on the total current value, as shown in Figure 3.1-8. The most common electronic signals are 4–20 milliamps DC, 10–50 milliamps DC, and 1–5 volts DC. Zero and span screws accessed through the transmitter housing provide adjustment capability for standard electronic transmitters, Figure 3.1-9.

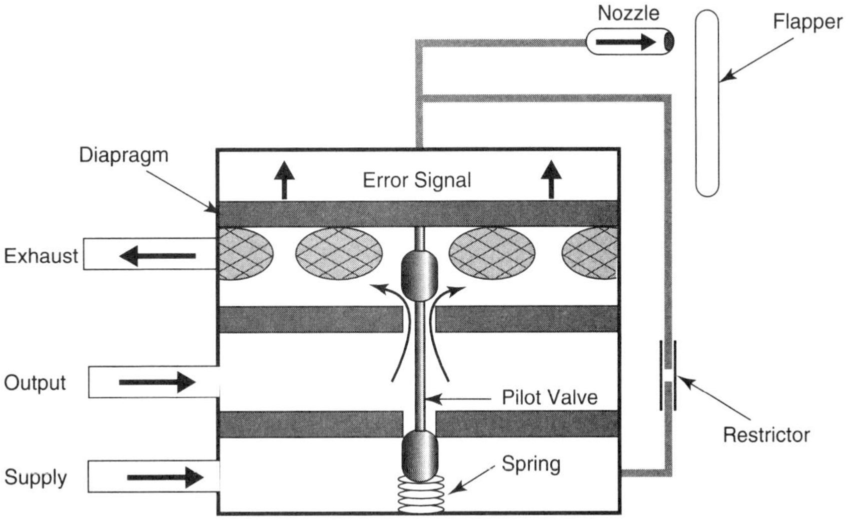

Figure 3.1-6 *Relay with Large Nozzle Gap*

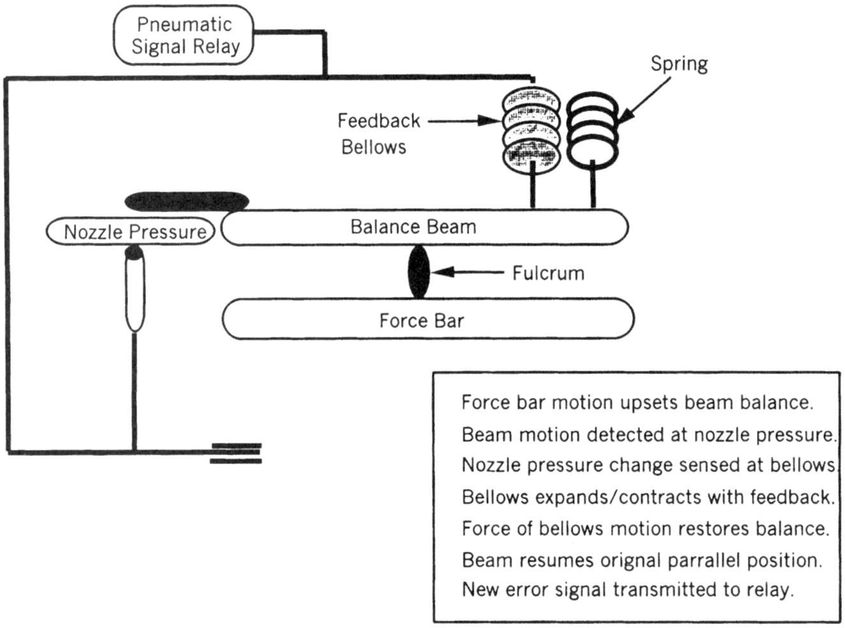

Figure 3.1-7 *Balance Beam Assembly*

Electronic transmitters require an external power supply to provide 24 to 48 VDC. The transmitter is connected to the power supply with two wires. The two-wire connection provides power to operate the transmitter, and also provides the propagation path for signal transmission. The transmitter varies the current flowing through the power/signal loop between 4 and 20 milliamps in relation to the process variable monitored. Transmitter circuitry consists of an input sensor (detector), connection board, calibration board, and an isolated output amplifier driver board. The detector converts physical displacement into a

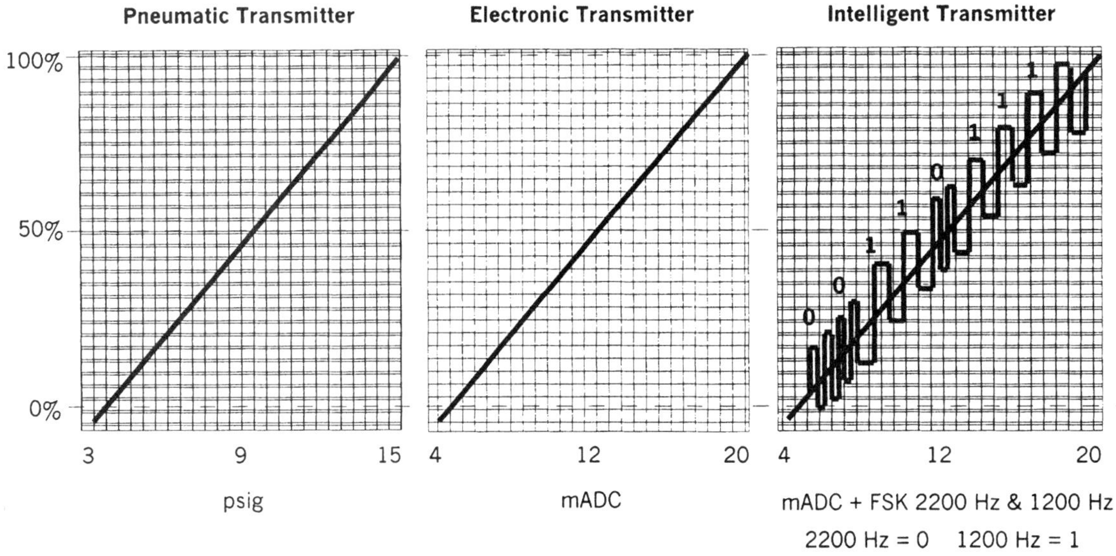

Figure 3.1-8 *Input to Output Signal Comparisons*

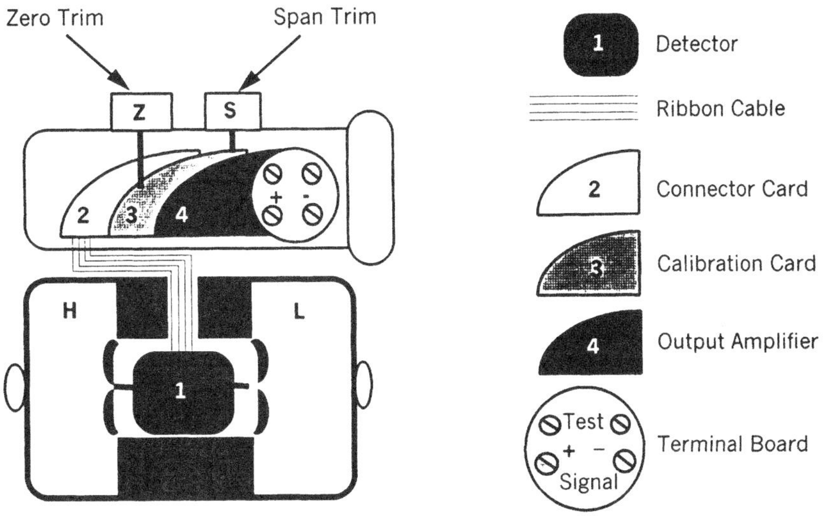

Figure 3.1-9 *Electronic transmitter Diagram*

change of impedance or capacitance. The connection board accepts the ribbon cable from the detector and conditions the signal for the calibration board. The calibration board consists of high impedance operational amplifier error detector circuits with tunable resistors (potentiometers). High impedance detection circuitry prevents the error detection amplifier from placing a load, (drawing current or dropping voltage), on the initial signal. The potentiometers are adjusted with the zero and span trim screws to correct calibration. The isolating output driver board converts the internal signal into a modulated current signal on the power loop wiring, Figure 3.1-10. The isolated output limits current on the loop and prevents the amplifier from damage in the event of a signal loop grounding condition.

Modern electronic transmitters are microprocessor based and have "intelligent" capabilities, Figure 3.1-12. They also require an external power source and modulate the loop current in relation to the process variable monitored. Intelligent transmitters monitor their own output signal loop for communication data. The binary communication data is superimposed on

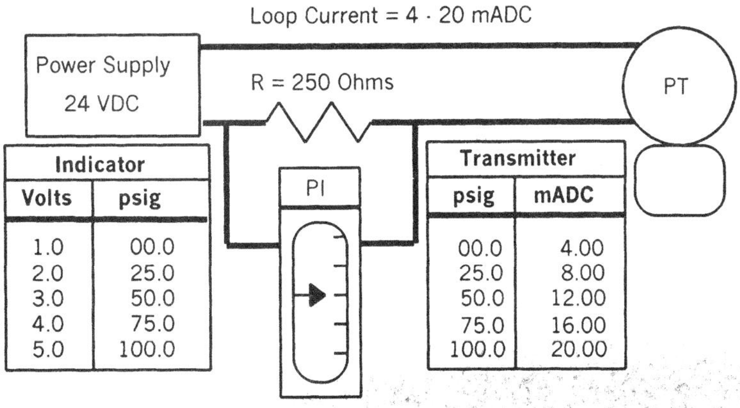

Figure 3.1-10 *Electronic transmitter Signal Loop*

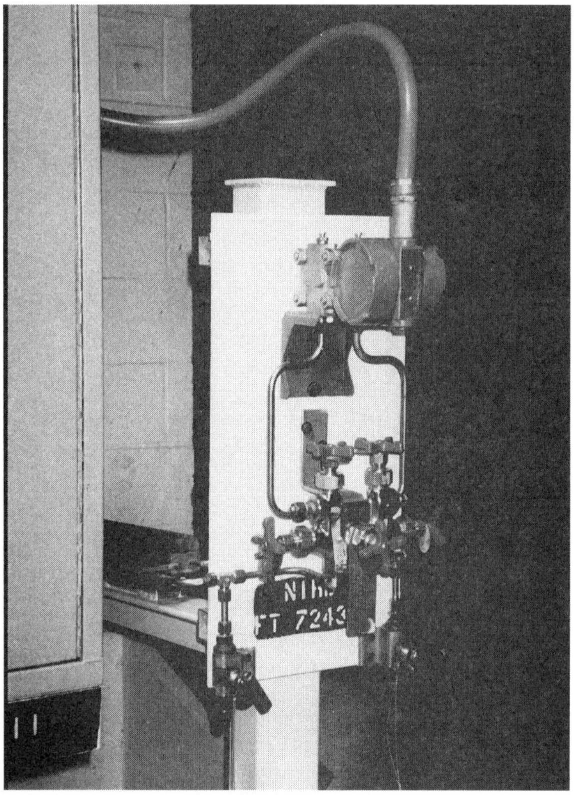

Figure 3.1-11 *Differential Transmitter*

the current signal as an AC frequency. Two distinct frequencies are used to represent the binary states. Typically 2200 Hertz represents a logic state (0) and 1200 Hertz represents a logic state (1). The AC frequency does not affect the current signal value, as it modulates equally above and below the current level and self-cancels. Since each transmitter is assigned a unique address, several transmitters often share a single two-wire power cable. This facilitates multi-drop and bus-type control arrangements where data from many transmitters is continuously monitored, displayed, and recorded by a central control station. Hand held communic ators and laptop computers access the intelligent transmitter software through the signal loop line to configure and calibrate the transmitter, Figure 3.1-13.

Pneumatic transducers (I/Ps) and (E/Ps) convert electronic signals into pneumatic signals. They contain many of the same type

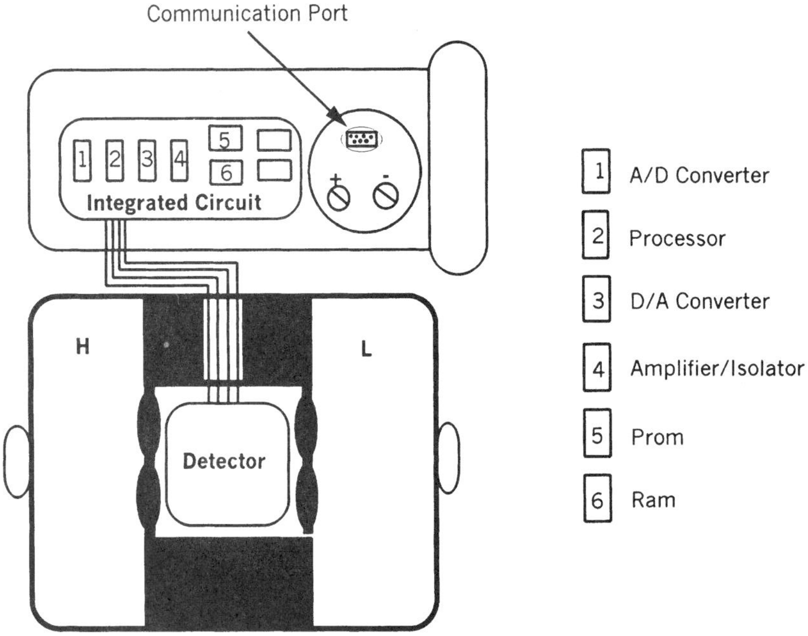

Figure 3.1-12 *Intelligent Transmitter*

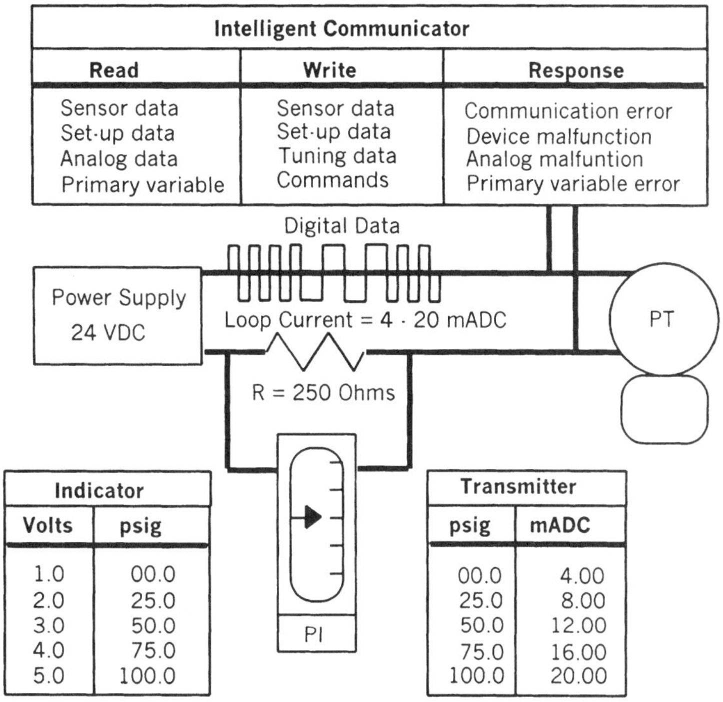

Figure 3.1-13 *Intelligent Transmitter Signal Loop*

components as pneumatic and electronic transmitters. The principle difference is, unlike transmitters, which detect a physical change in process energy, transducers merely convert one form of signal transmission data into another form of signal transmission data. The current to pneumatic transducer (I/P) contains a pneumatic relay, a flapper/nozzle assembly, a balance beam, a feed-back bellows, and a coil. The operating principle of the pneumatic

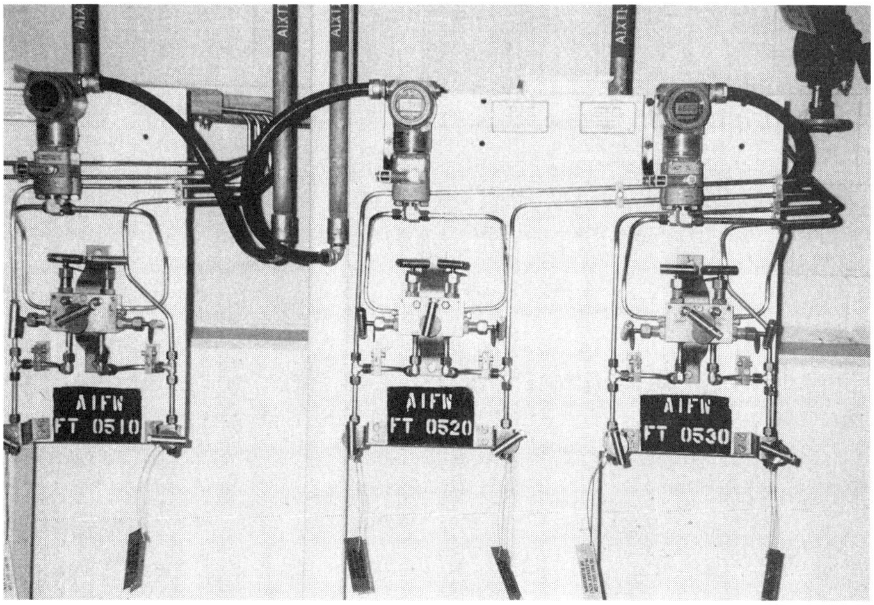

Figure 3.1-14 *Intelligent D/P Transmitters*

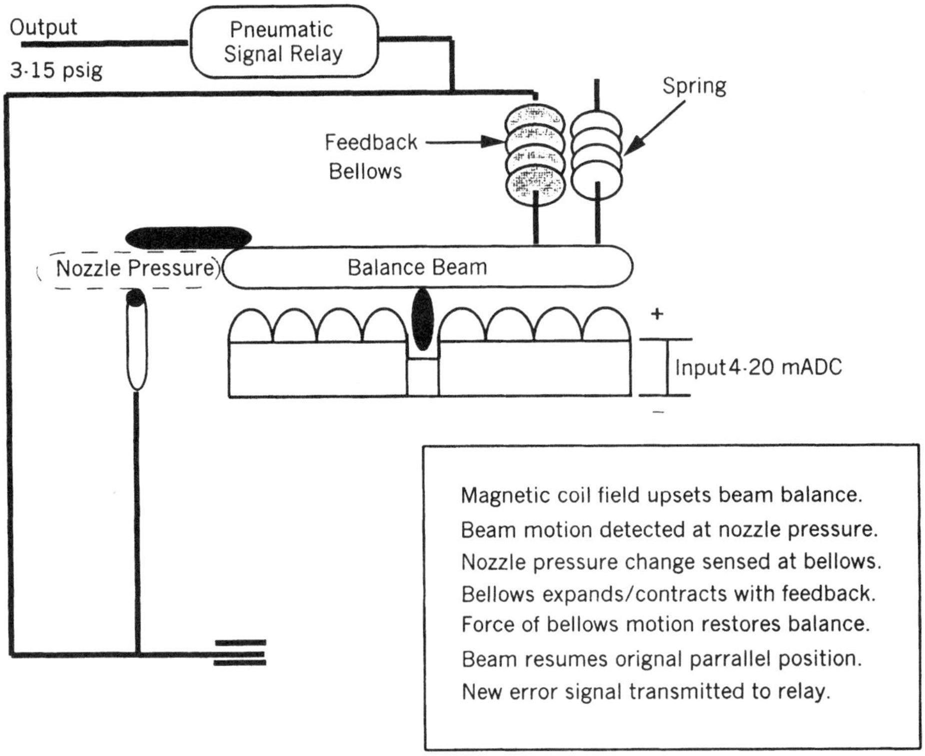

Figure 3.1-15 *Electro/Pneumatic Balance Beam Assembly*

section is identical to that of the pneumatic transmitter. While the pneumatic transmitter relies on a force bar to initiate the unbalance condition to the balance beam, the I/P depends on an electrical coil to disrupt beam balance in proportion to the electrical current applied to it, see Figure 3.1-15. Typical current inputs are 10–50 milliamps DC and 4–20 milliamps DC. Typical outputs are 3–15 psig and 6–30 psig.

3.2 Physical Displacement Transducers

Physical Displacement transducers convert physical displacement (motion) into an electronic signal or an electric signal into physical displacement. The fundamentals of solenoid, coil, and core behavior are observed as two magnets placed poles opposed create a repelling force between them sufficient to generate a physical displacement. Figure 3.2-1 illustrates a ferrous material placed in a magnetic field increasing the field strength, and a magnetic field applied to a ferrous material inducing a force. A complete explanation of electro/mechanical relationships controlling coil and core dynamics is beyond the scope of this text. Exact detail of these phenomena is found in the works of Michael Faraday and Emil Lenz.

Common physical displacement transducers include the solenoid, LVDT (linear variable displacement transformer), the eddy current detector, the seismic detector, and the velocity

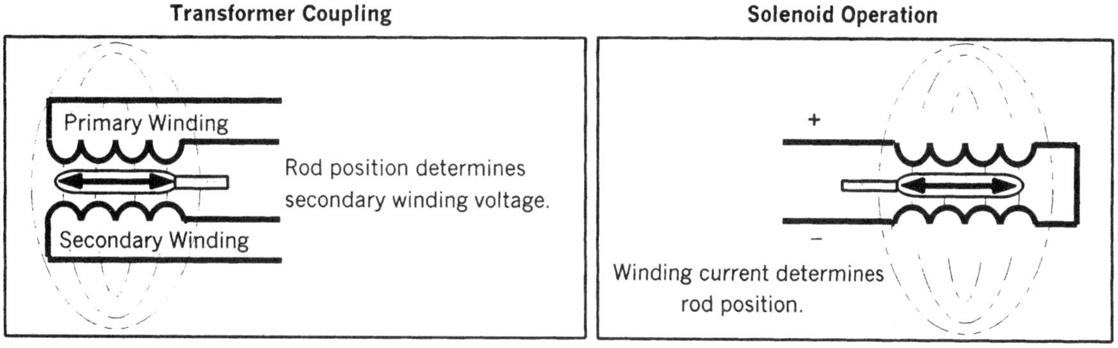

Figure 3.2-1 *Solenoid, Coil, Core Operation*

transducer. Solenoids convert electric current into physical displacement. They consist of a primary coil winding and a movable core (slug). The core is often attached to small three way valves, but is also connected to strikers and plungers. Solenoids attached to small valves are often used to direct fluid for control purposes. Solenoid valves are also installed in the instrument air supply line to air-operated valves to open or close the valve in response to process conditions. Solenoids attached to strikers and plungers lock and unlock doors, push small items from conveyors, and perform a multitude of other functions where a small displacement device is electrically controlled. A relay is a set of switching contacts manufactured with an integral internal solenoid. The configuration depicted in Figure 3.2-2 utilizes the relay coil (solenoid) to pull-in the switch contacts. Relays are used throughout industry for remote automatic switching of electrical devices. Solenoids and relays are the main control initiating devices of program logic functions.

The linear variable displacement transformer (LVDT) converts linear displacement into an equivalent change in voltage. Their most common application is as position feed-back transmitters. The LVDT is installed on a valve stem or mechanical lever. Any change in the position of the valve stem or lever is transmitted back to a control system such as a servo control assembly. The servo control assembly compares desired position to actual position

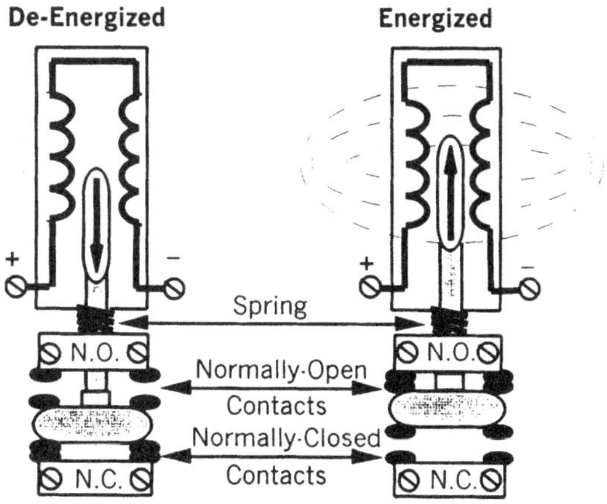

Figure 3.2-2 *Relay/Solenoid*

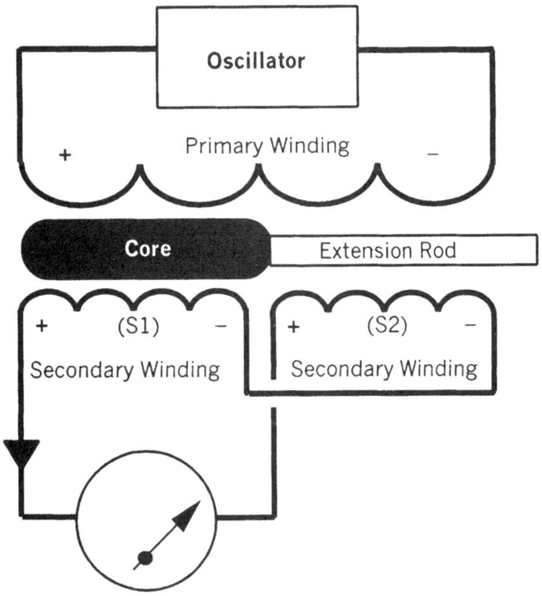

Figure 3.2-3 *LVDT at Max Positive Position*

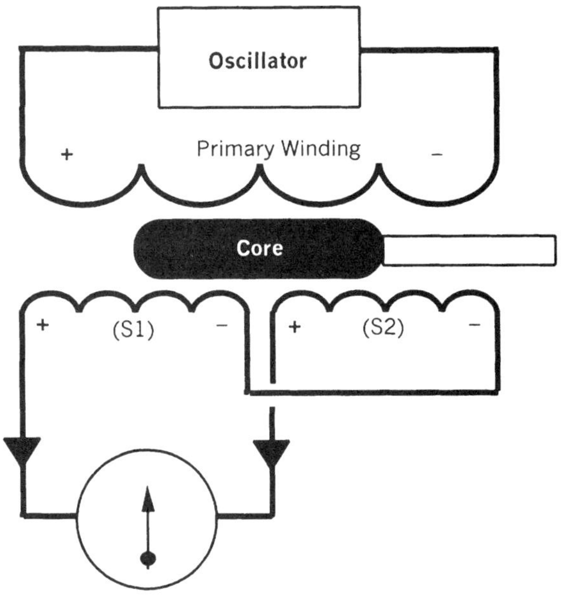

Figure 3.2-4 *LVDT at Null/Center Position*

and develops a correction signal to correct any deviation. The LVDT assembly consists of an excitation oscillator, a primary transformer, a secondary transformer, and a movable core. The excitation oscillator generates an alternating voltage to the primary transformer. The primary transformer generates a flux field to the secondary transformer. The secondary transformer consists of two separate coils wired series opposing. The position of the movable core determines the amount of coupling strength to the secondary coils. As the secondary transformer is series opposed, and increasing coupling to one coil decreases coupling to the other, voltages range from positive polarity saturation to negative polarity saturation with the exact center self canceling. Figures 3.2-3, 3.2-4, and 3.2-5 show each position. The center travel position where the voltages sum to zero is the null position. LVDTs are produced to detect a wide range of values. The outputs are always bipolar (going positive and negative) with zero volts at the null or center position.

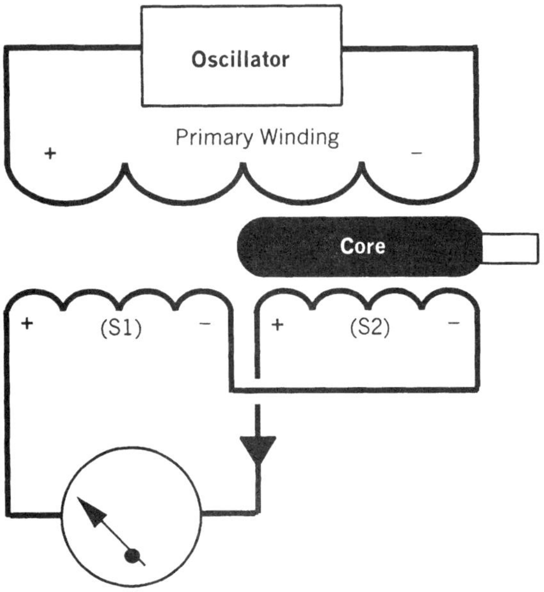

Figure 3.2-5 *LVDT at Max Negative Position*

Eddy current detectors convert mils gap into voltage. Mils gap is the distance between the probe tip and the metallic object monitored

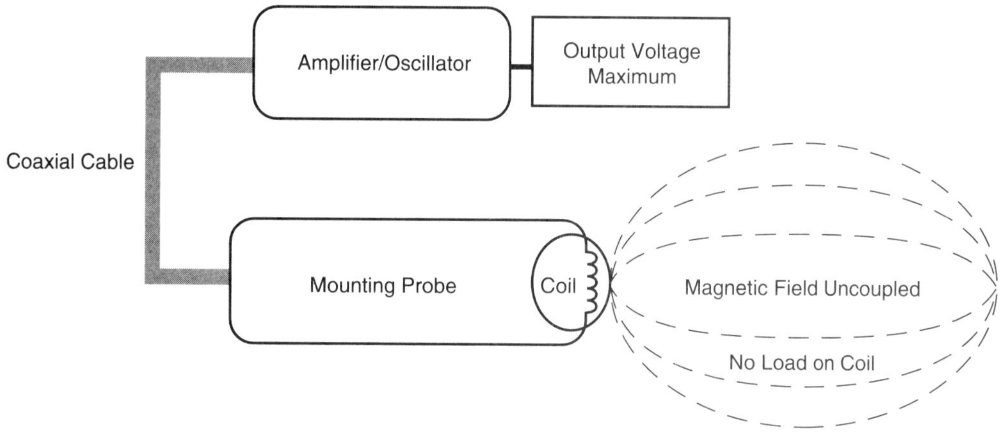

Figure 3.2-6 *Proximity Detector with No Load*

measured in one-one thousandths of an inch, Figures 3.2-6 and 3.2-7. An excitation oscillator provides a high frequency voltage to the detector coil. The energized coil emits a magnetic field. The strength of the magnetic field changes as metallic mass passes through close proximity. The change in field strength is detected and converted into an equivalent voltage value (- 0.200 VDC per mil). Eddy detectors are most often used as non-contacting vibration detectors on rotating equipment. Typical input range values are 0–100 mils producing outputs of 0–(-)20 VDC.

Seismic detectors convert gravity into volts. The components of seismic detectors include a strain gauge or piezoelectric crystal and an amplifier. Seismic detectors are used to detect ultra high frequency vibration. Velocity detectors convert acceleration into volts. A movable suspended core inside of an excited coil detects velocity. An amplifier provides the excitation current to the coil. The suspended core moves in response to accelerating forces. The output

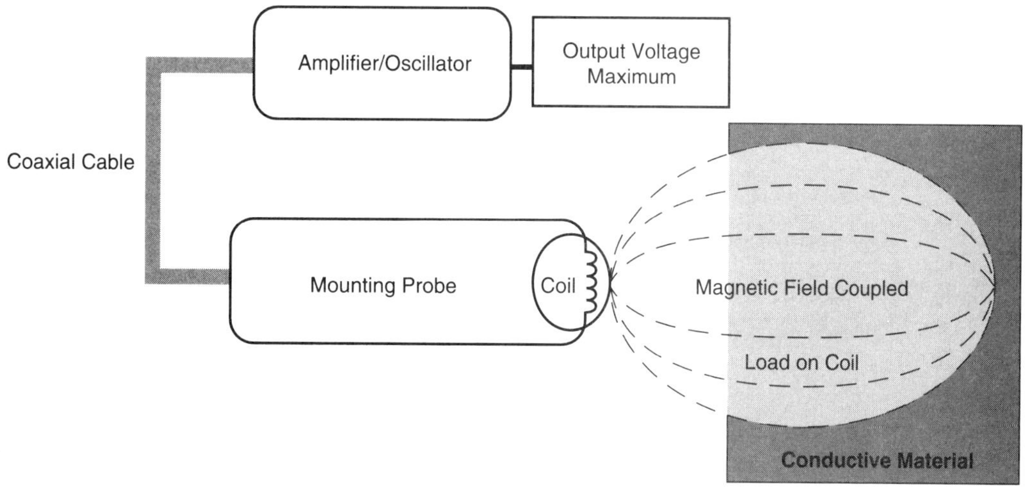

Figure 3.2-7 *Proximity Detector with Load*

voltage is generated as the movable core cuts lines of coil flux. The rate at which the lines are cut determines the amplitude of the voltage generated.

3.3 Electronic Transducers

Electronic transducers are generally not used as stand alone instruments. They are usually integrated with other components to detect outputs of thermocouples, RTD, thermistors, mass flow detectors, and analytical cells. Investigation of electronic transducers requires the application of several fundamental laws of electricity. The common units are volts, amperes, and ohms. Voltage, (E), refers to the potential energy available and is sometimes considered analogous to the pressure of a fluid. Amperes, (current or I), refers to the actual quantity of electrons per second flowing past a fixed point. Resistance (Ohms or R) is the amount of opposition to current flow exhibited by a circuit or device. Resistors in series circuits create an opposition force equivalent to the sum of the individual resistances. Resistors in parallel circuits exhibit an opposition force equivalent to the reciprocal of, the sum of the reciprocals of, the individual resistors. Examples are given in Figures 3.3-1 and 3.3-2.

The first basic (Ohms law) is actually a statement of equilibrium. The fundamental equilibrium of an electric circuit is stated as E ÷ I x R = 1, or E = I x R (voltage is equal to the product of current and resistance). To determine the unknown values substitute the known values into the statement and reduce. The second principle is Kirchoff's statement of conservation. It states that the sum of the voltage drops across a circuit are equal to the total voltage applied to the circuit, and that the amount of current entering a circuit must equal the amount of current leaving the circuit. The third principle is Thevenin's theorem, which states that any combination of series and parallel components in a circuit can be represented by a single equivalent voltage and resistance, as in Figures 3.3-3 and 3.3-4.

There are many different applications of electronic detection transducers. Common

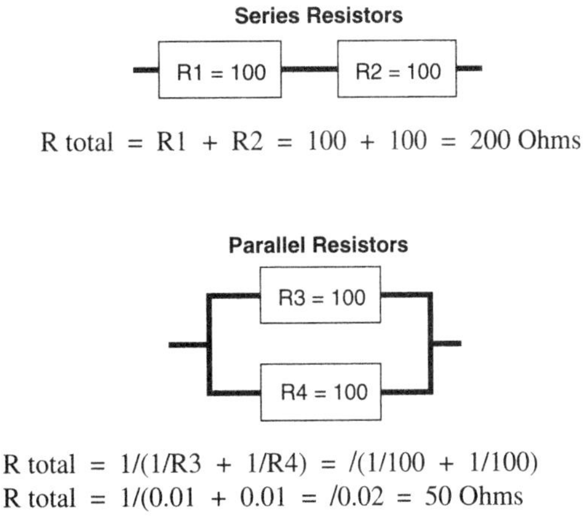

Figure 3.3-1 *Simple Series and Parallel Resistance*

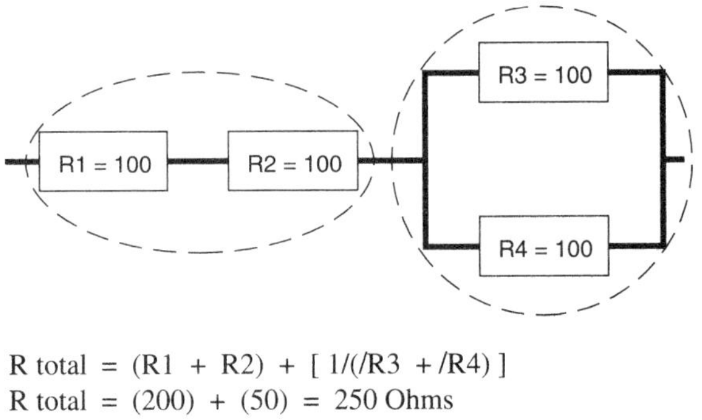

Figure 3.3-2 *Combined Series and Parallel Resistance*

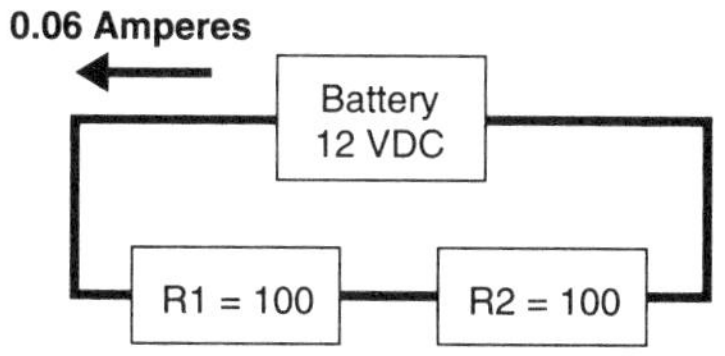

Ohms equilibrium statement E/I × R = 1
Substitute knowns 12/I × 200 = 1
Solve for unknown 12/200 = I = 0.06 Amperes

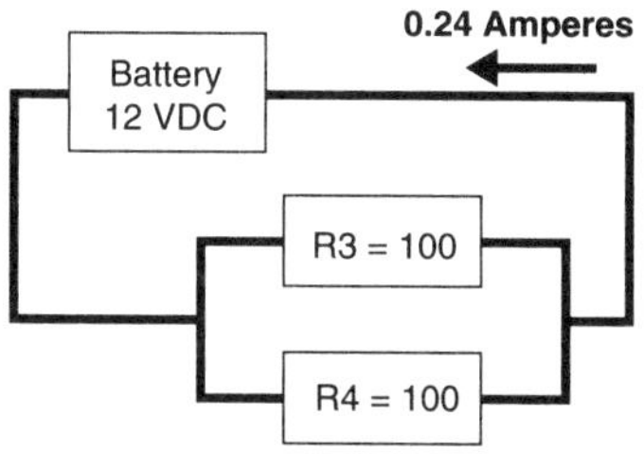

Ohms equilibrium statement E/I × R = 1
Substitute knowns 12/I × 50 = 1
Solve for unknown 12/50 = I = 0.24 Amperes

Figure 3.3-3 *Simple Series/Parallel Circuit*

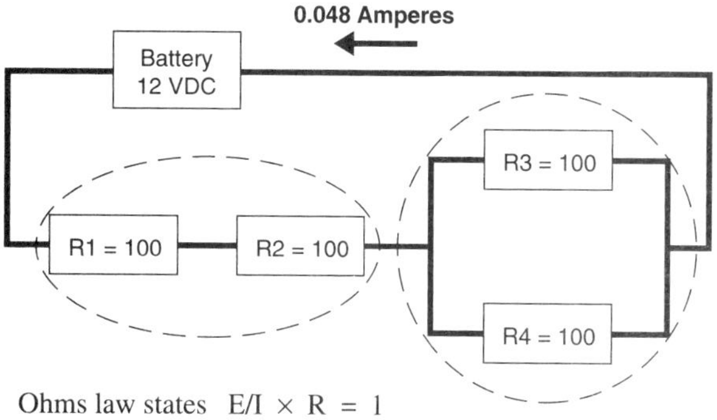

Ohms law states E/I × R = 1
12/I × 250 = 1
12/50 = I
0.048 = I = 0.048 Amperes

Figure 3.3-4 *Combination Series/Parallel*

elements include a bridge circuit and an amplifier section. The most basic designs also include a De'Arsonval meter or galvanometer. The bridge circuit is the building block of micro-detection circuitry. The bridge is a configuration of parallel precision resistors arranged in such a manner as to detect the minutest changes in current between them. Once the bridge is balanced (nulled), the monitored component is placed into the circuit. Any deviations induced are attributed directly to changes in the monitored component. Figures 3.3-5, 3.3-6, and 3.3-7 depict three

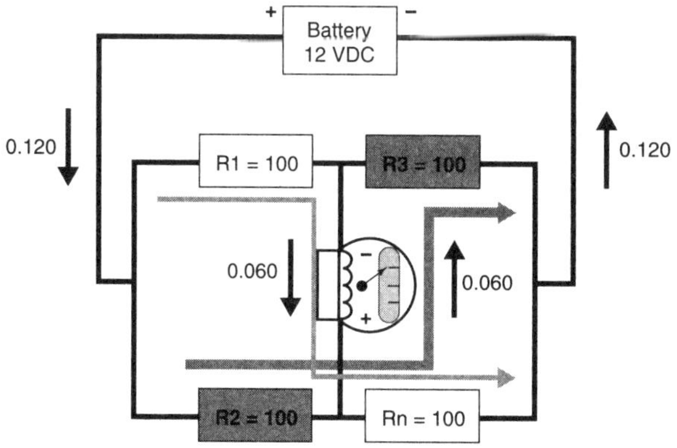

Amperes = Volts/Ohms
I R1/Rn = 12/200 = 0.060
I R2/R3 = 12/200 = 0.060
I Meter = I R1/Rn - IR2/R3 = 0.060 - 0.060 = 0 = Null

Figure 3.3-5 *Bridge Nulled for 100 Ohms*

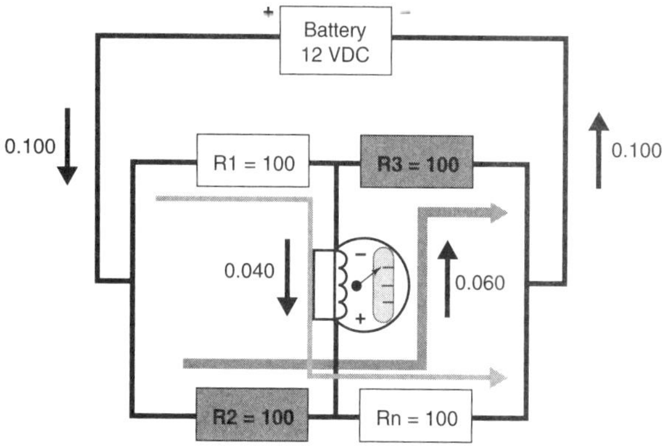

Amperes = Volts/Ohms
I R1/Rn = 12/300 = 0.040
I R2/R3 = 12/200 = 0.060
I Meter = I R1/Rn - IR2/R3 = 0.040 - 0.060 = (-)0.020

Figure 3.3-6 *Bridge Detecting 200 Ohms*

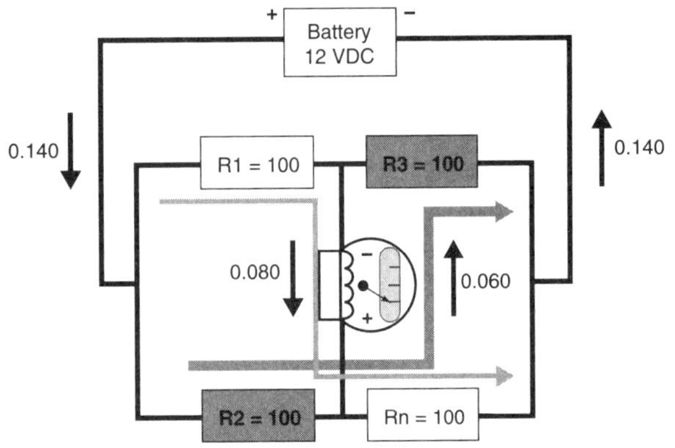

Amperes = Volts/Ohms
I R1/Rn = 12/150 = 0.080
I R2/R3 = 12/200 = 0.060
I Meter = I R1/Rn - I R2/R3 = 0.080 - 0.060 = 0.020

Figure 3.3-7 *Bridge Detecting 50 Ohms*

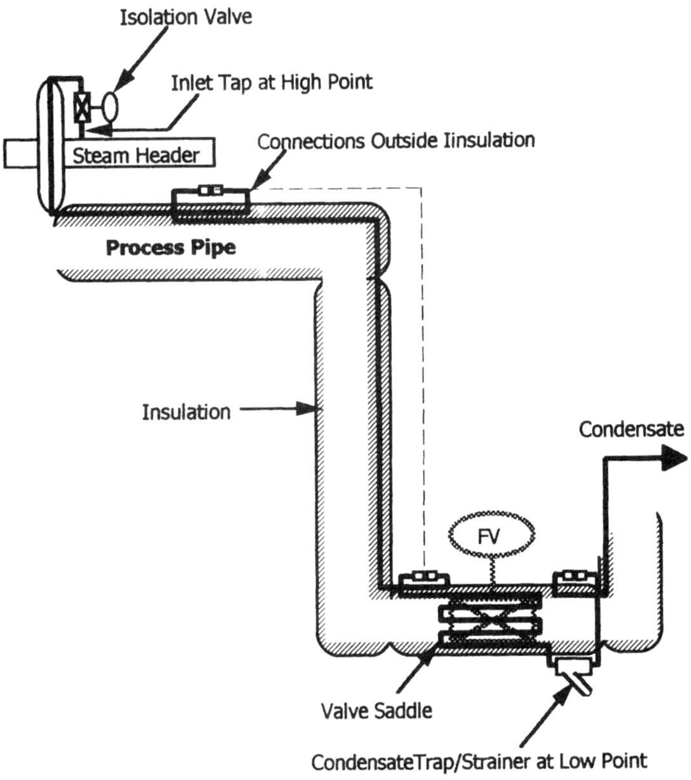

Figure 3.4-1 *Steam Trace*

different conditions of bridge bias. Bridge circuits detect deviations in temperature, flow, and analytical cells.

3.4 Heat Trace

Heat tracing for transmitters and impulse lines prevents fluids in the lines from freezing, and in some instances prevents particular process medias from cooling to a solid state. Steam and electric heat cable are used as heat tracing. Steam systems transfer heat to the sensing lines by providing steam flow through adjacent tubing. As steam relinquishes heat to the sense lines, it cools and condenses. Condensation traps are installed at various low points in the steam tubing runs to collect moisture, Figure 3.4-1. Steam traps are designed to vent accumulated condensate when it reaches a predetermined level in the trap. Most steam trace performance problems are related to steam trap malfunctions, Figure 3.4-2. A drip pot that fails to blow-down forces condensate back into the steam heat tubing and stops steam flow.

Electric heat trace systems use heat strip tape to transfer heat to sense lines. Heat tape is a semi-conductive wire pair. It looks like a rigid extension cord, and is produced in various lengths and power ratings. Various methods are used to energize the heat strip. More complex

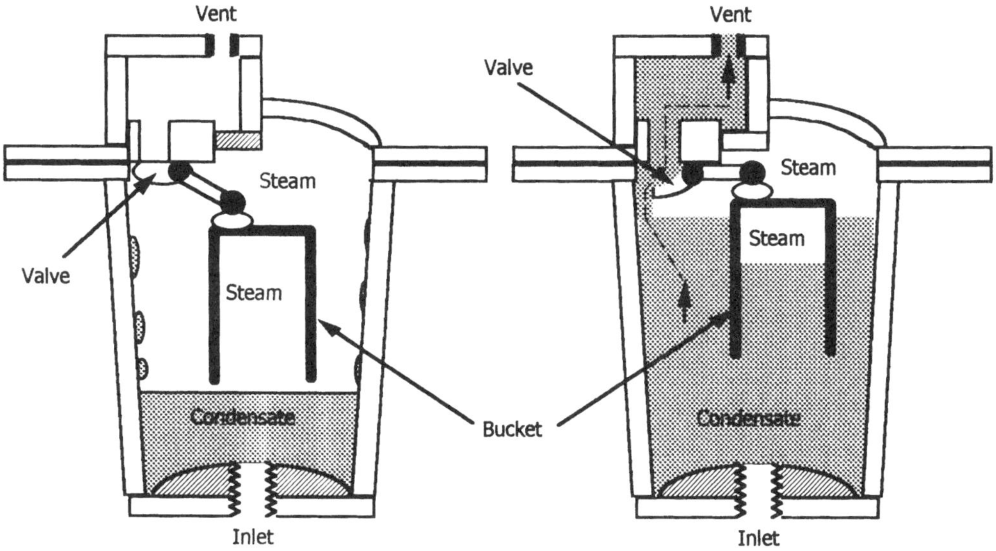

Figure 3.4-2 *Bucket Steam Trap*

systems feature a temperature control unit that monitors the temperature of each individual strip and adjust heating current as required to maintain a preset temperature, Figure 3.4-3. A separate module with status and diagnostic LEDs, light-emitting diodes provides

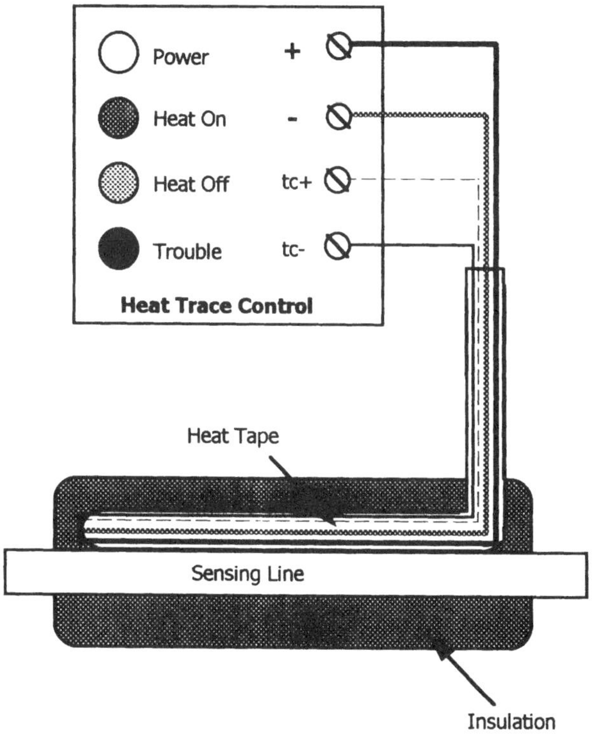

Figure 3.4-3 *Electric Heat Trace*

Figure 3.4-4 *Electric Heat Trace Control Panel*

power and indication for each section of heat strip. Heat tape degrades with time, and, as with any power supply, modules occasionally fail. Continuity testing generally determines the condition of the heat strip. Module failure is readily tested by component substitution.

Chapter Three Summary:

- A transducer is a device that converts one form of signal data into another proportionally equivalent form of signal data. A transmitter is a device that conditions and conveys signal data.Transmitters detect and convert changes in pressure, flow, or temperature into equivalent pneumatic or electronic signals and transmit the signal data from the field to a control center instrument (controller, recorder, or indicator).
- The measured variable value from the primary element is applied to the input sensor (input). The input sensor transfers an equivalent amount of energy to a force balance mechanism.
- A flapper nozzle assembly detects the amount of deviation at the force balance vbeam. The flapper nozzle assembly backpressure is amplified by a pneumatic relay and retransmitted as a pneumatic output signal.
- Electronic transmitters produce an electrical output signal.
- Intelligent transmitters monitor their own output signal loop for communication data.
- Hand held communicators and laptop computers access the intelligent transmitter software through the signal loop line to configure and calibrate the transmitter.
- I/P and E/P transducers convert electrical signals into equivalent pneumatic signals. They are commonly used in the field to supply instrument air to final control elements.
- Physical displacement transducers convert physical displacement (motion) into an electronic signal or an electric signal into physical displacement.
- Common physical displacement transducers include the solenoid, LVDT (linear variable displacement transformer), the eddy current detector, the seismic detector, and the velocity transducer.
- Solenoids convert electric current into physical displacement.
- A relay is a set of switching contacts manufactured with an integral internal solenoid.
- Relays are used throughout industry for remote automatic switching of lectrical devices.
- Solenoids and relays are the main control initiating devices of program logic functions.
- The LVDT converts linear displacement into an equivalent change in voltage. Their most common application is as position feed-back transmitters. The center travel position where the voltages sum to zero is the null position. The outputs are always bipolar (going positive and negative) with zero volts at the null or center position.
- Eddy current detectors convert mils gap into voltage. Mils gap is the distance between the probe tip and the metallic object monitored measured in one-one thousandths of an inch. Eddy detectors are most often used as non-contacting vibration detectors on rotating equipment.

- Seismic detectors convert gravity into volts. Seismic detectors are used to detect ultra high frequency vibration.
- Velocity detectors convert acceleration into volts.
- Electronic transducers are generally integrated with other components to detect outputs of thermocouples, RTDs, thermistors, mass flow detectors, and analytical cells.
- Resistors in series circuits create an opposition force equivalent to the sum of the individual resistances.
- Resistors in parallel circuits exhibit an opposition force equivalent to the reciprocal of, the sum of the reciprocals of, the individual resistors.
- Ohms law states the fundamental balance in any electric circuit is stated as $E/I \times R = 1$
- Kirchoff's statement of conservation is the sum of the voltage drops across a circuit are equal to the total voltage applied to the circuit, and that the amount of current entering a circuit must equal the amount of current leaving the circuit.
- Thevenin's theorem states that any combination of series and parallel circuits can be represented by a single voltage and resistance.
- Common elements of electronic detection transducers include a bridge circuit and an amplifier section.
- A Wheatstone bridge is a configuration of parallel precision resistors arranged in such a manner as to detect the minutest changes in current between them.
- A balanced bridge is (nulled). Bridge circuits detect deviations in temperature, flow, and analytical cells.
- Heat tracing for transmitters and impulse lines prevents fluids in the lines from freezing, and in some instances prevents particular process medias from cooling to a solid state.
- Steam and electric heat cable are used as heat tracing.
- Steam traps are installed at various low points in the steam tubing runs to remove condensate from the system.

Chapter Two Review Questions:

1. Define transducers and transmitters.
2. Describe function of a pneumatic transmitter.
3. State the purpose of the input sensor.
4. State the purpose of the force balance mechanism.
5. State the purpose of the output relay.
6. Describe the function of an electronic transmitter.
7. State the purpose of the input sensor.
8. State the purpose of high impedance and isolation.
9. State the purpose of the error detector.
10. State the purpose of the amplifier/driver.
11. Contrast isolated and un-isolated outputs.
12. State the purpose of physical displacement transducers.
13. Describe the operation of a solenoid.
14. Describe the operation of an electrical-switching relay.
15. State the function of an LVDT (inches/VDC).
16. List the major functional components of an LVDT.
17. State the function of an eddy current proximeter (mil gap/VDC).
18. State the function of a seismic detector (G/volts).
19. State the function of a velocity detector (Acceleration/volts).
20. State Ohm's law of equilibrium.
21. State Kirchoff's law of conservation.
22. Calculate the combined resistance of a series circuit.
23. Calculate the combined resistance of a parallel circuit.
24. Calculate the combined resistance of a series/parallel circuit.
25. Develop a diagram of a simple bridge circuit.
26. State two methods of freeze-protecting sense lines.

Chapter 4

Analyzers

OBJECTIVES

Upon completion of Chapter Four the student will be able to:

- *State the definition of chemical analysis.*
- *State the purpose of qualitative analysis.*
- *State the purpose of quantitative analysis.*
- *List several industrial applications of chemical analysis.*
- *State the purpose of analytical instruments.*
- *State the undesirable effect of introducing oxygen into some processes.*
- *List three types of detectors that comprise common analytical probe assemblies.*
- *State the purpose of a measurement cell.*
- *State the purpose of a reference cell.*
- *State the purpose of a temperature compensation detector.*
- *Explain the relationship between potential difference and solution concentration.*
- *State the purpose of pH analyzers.*
- *Identify the acidity or alkalinity of a substance based on it's pH value.*
- *State the purpose of oxygen analyzers.*
- *State the purpose of oxygen permeable membrane in an oxygen probe.*
- *State two common oxygen analyzer malfunctions.*
- *State the purpose of conductivity analyzers.*

- *Describe the relationship between conductivity (S) and resistance.*
- *State two terms used for conductivity.*
- *State the definition of specific conductance, (K).*
- *List several sample criteria required to produce precise analytical values.*
- *State the most reliable method to determine operability of an analyzer probe.*
- *State the adjustment used to correlate an analyzer reading to a known sample.*
- *Describe the troubleshooting technique used to isolate a reference temperature malfunction.*
- *Define adsorption.*
- *Explain the application of adsorption as used in chromatographic analysis.*
- *Define chromatogram as used in chromatography.*
- *State the purpose of chromatography as used in industrial applications.*
- *List three steps of the chromatographic process.*
- *State the purpose of a sampler valve.*
- *State the purpose of the mobile phase.*
- *List several carrier gasses.*
- *State the purpose of the stationary packing.*
- *List several examples of stationary column packing.*
- *Describe the relationship between absorption and sample transmission time.*
- *State the purpose of the spectrometer as it applies to chromatography.*
- *List four functional sections of a spectrometer.*
- *Describe the function of a spectrometer.*
- *Explain the relationship between travel time and concentration.*
- *List several potential chromatograph malfunctions.*
- *List the significant sample parameters controlled in chromatographic analysis.*
- *State the primary discriminating sample variable in chromatographic analysis.*
- *Describe the distillation process.*
- *Describe the fractional distillation process.*
- *Differentiate between simple distillation and fractional distillation.*
- *Describe the purpose of chromatographs in the fractioning process.*
- *Explain trace detection as used by chromatographs in control applications.*

4.0 Introduction

In most process plants analytical tests are continuously performed to determine the quality of raw materials, intermediate, and finished products of the unit. Most of these tests are in the form of chemical analysis. Chemical analysis is the science and technology used to identify and measure the chemical composition of a product. Qualitative analysis determines the presence of a particular substance in a sample. A test to determine whether a sample contains CO_2 or that the sample has a certain pH is an example of qualitative analysis. Quantitative analysis determines the amount of a particular substance in a sample. Such determinations as the percent of oxygen or the "parts per million" of a contaminate are examples of quantitative analysis. Some analyzers, such as a gas chromatograph, perform both qualitative and quantitative analysis. Measured values are expressed as percent of concentration, parts per million (PPM), parts per billion (PPB), moles, seimens, micro Mhos, and pH. Common industrial applications of chemical analysis include monitoring for product purity, verifying concentration of solutions, and detecting unwanted or undesirable contaminates. Industries maintain chemical laboratories to support skilled technicians and chemists that routinely pull and perform detailed laboratory analysis of product samples. On-stream analytical instruments are installed to perform continuous on-line monitoring, recording, control, and annunciation of critical phases of an operation. This helps to assure in-process product quality and reduces chances for environmental pollution. Most on-stream analyzers work on the same principles as their laboratory counterparts, excepting the addition of mechanisms and circuitry required to support unattended automatic sampling. Process analyzers may be classified in various ways depending on the purpose of the classification. Classification can be (a) by operating principle (infrared, ultraviolet, chromatographic, etc.), (b) by type of analysis (pH, oxygen, carbon dioxide, etc.), and (c) by selection (selective or non-selective). Infrared may be sensitized to monitor a single component while a chromatograph may monitor several components.

The number of different analyzers employed in the process industries makes up a significant list. They include pH analyzers, gas chromatographs, mass spectrometers, colorimeters, viscometers, density/specific gravity, sulfur analyzers, oxygen analyzers, opacity, conductivity, combustible gas, and turbidity. There are many others developed specifically for refinery streams such as ASTM distillation, pour point, flash point, and octane, etc. In some applications, analytical instruments monitor the chemical component of the process variable and provide control action. Analyzer control loops are common in chemical feed systems. Percent concentration of a particular chemical is monitored as the process variable. Sample concentration is compared to set point, and the analytical controller outputs a demand signal to the chemical feed pumps that increases or decreases the amount of chemical injected with each stroke of the pump. Many processes are extremely sensitive to oxygen. Unstable compounds can exert tremendous energy if allowed to oxidize, and combustible processes can spontaneously ignite if exposed to oxygen. Oxygen analyzers are often installed to reduce corrosion and combustibility in these processes. This text will review in detail five analyzers: pH, conductivity, oxygen, spectrometers, and gas chromatographs.

4.1 pH Analyzers

pH monitoring analyzers measure the alkalinity or acidity of an aqueous solution. pH is actually a measurement of hydrogen ion concentration. Acidic solutions contain more hydrogen ions than base solutions. pH is actually the negative logarithm of hydrogen ion concentration. Therefore, a change of one unit pH is equivalent to a ten-fold change in the effective strength of an acid or base. Highly acidic concentrations are referenced at (0) on the pH scale. Pure water is neutral (7), and highly alkaline (base) concentrations are referenced at (14). Figure 4.1-1 demonstrates the pH of several common products. Since pH measurement is temperature sensitive, pH analyzers contain a temperature-sensing element. pH probes are actually a combination of three detectors; a measurement cell, a reference cell, and a temperature compensation detector. The most common type of pH sensing element consists of a glass measurement electrode, a calomel reference electrode, and a temperature compensation RTD, as shown in Figure 4.1-2. Measurement cells detect the concentration or activity of the sample, reference cells generate a constant fixed reference voltage, and temperature compensators provide a temperature feed-back signal. Measurement and reference cells function on principles similar to a battery. Conductive elements are selected to react with a specific electrolytic solution. As the conductor is placed in solution, a chemical reaction occurs. This reaction results in an ionic transfer between the solution and the conduc-

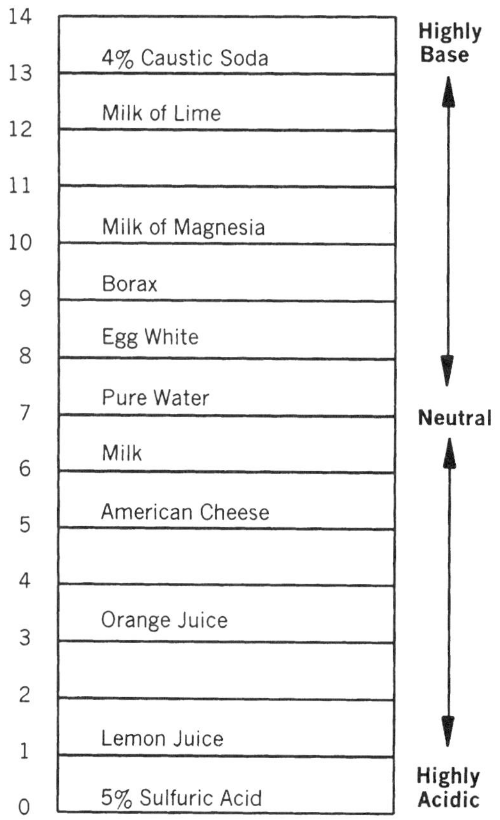

Figure 4.1-1 *The pH Scale*

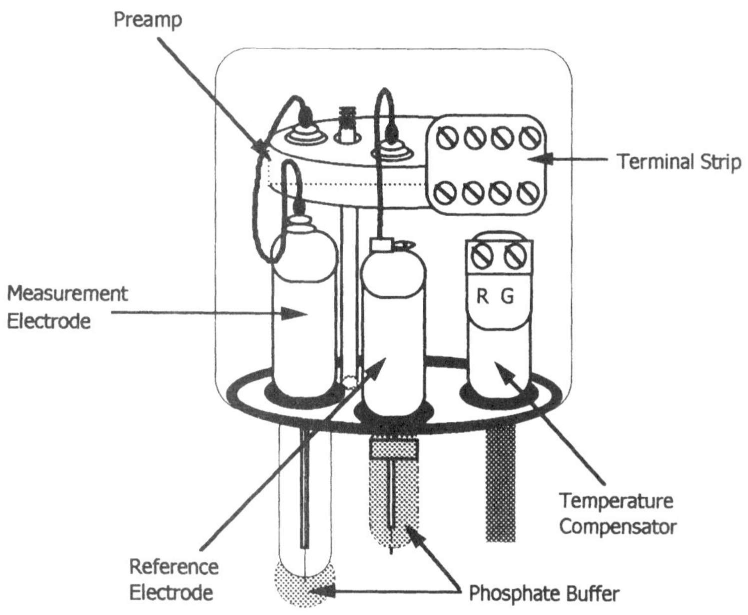

Figure 4.1-2 *Analytical Probe Assembly*

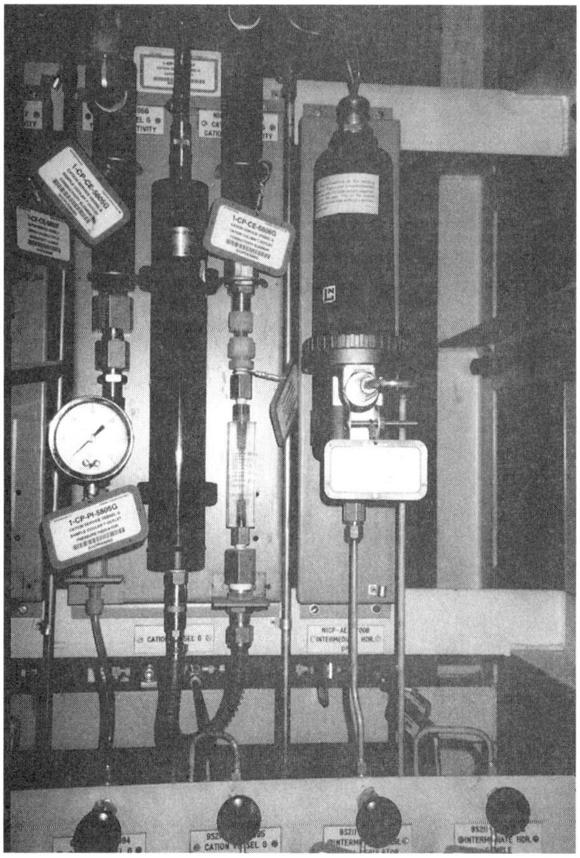

Figure 4.1-3 *pH Probe Assembly*

tor. Anode elements develop a positive charge, and cathode elements develop a negative charge. The potential difference between charges is proportional to the logarithm of the analytic solution concentration. pH reactions are temperature sensitive and require accurate temperature-sensing feedback. An RTD circuit placed in the sample solution at the same location as the detection cells develops a temperature related resistance value. Electronic circuits in the analyzer use the RTD signal to develop a temperature compensation logarithm. This logarithm is applied to the cell signal to develop a temperature-corrected value. Probe-generated voltage values are inherently diminutive in magnitude. Bridge circuits located in the analyzer detect the minute changes in cell signal values. The signals are then amplified and processed as normal electronic data. Most pH analyzers provide local and remote indication, local and remote annunciation, and controlled output capability. Output indication is scaled as 0–14 pH.

4.2 Oxygen Analyzers

Two common oxygen analyzers are the paramagnetic phototube detector and the galvanic probe detector. Paramagnetic analyzers detect the relative magnetic susceptibility of gasses. Oxygen has a strong attraction to magnetic fields while most other gasses have either a weak attraction or repulsion to magnetism. Figure 4.2-1 shows the magnetic susceptibility of several gases. Relationship phototube detectors consist of a light source, a test body/mirror assembly, a beam splitting mirror, two phototube/optical amplifiers, a detecting bridge, and the supporting electronic amplifiers and signal conditioning circuits, as in Figures 4.2-2 and 4.2-3. The test body/mirror assembly is suspended in the test chamber and subjected to a magnetic field. A light source is directed at the test mirror. The test mirror directs the light beam to a beam splitting mirror. Under null conditions the beam splitter divides the light equally between the phototubes so their resultant voltage output is equal. As the sample gas is introduced into the test chamber, the test body deflects proportional to the magnetic susceptibility of the sample gas. The attached mirror rotates with the test body and redirects the light distribution to the beam splitter. Because phototubes generate a voltage output equivalent to the amount of light directed to them, the difference in output between the phototubes is proportional to the magnetically induced deflection of the detector mirror. This difference is detected and amplified by the bridge and amplifier section and conditioned into a suitable readout. Operability checks are performed by admitting normal atmosphere into the test body and observing an indication of 20.7%.

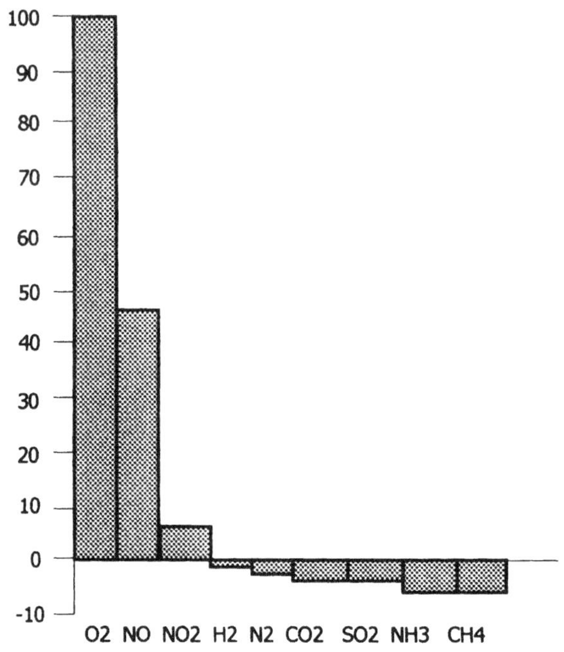

Figure 4.2-1 *Relative Magnetic Susceptibility of Gases*

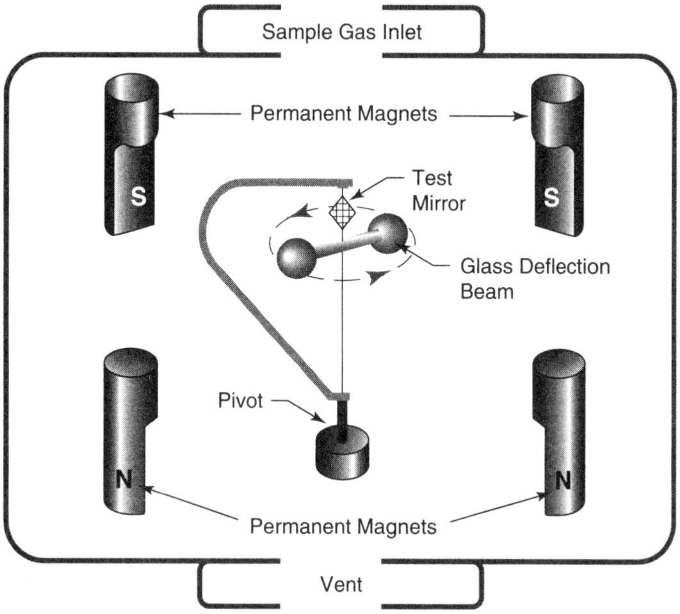

Figure 4.2-2 *Paramagnetic Detector*

Electrolytic probes often detect dissolved oxygen in fluids. Electrolytic oxygen analyzers use a platinum micro-cathode measuring electrode and a silver chloride macro-anode reference electrode placed in a phosphate buffer solution. An oxygen permeable membrane isolates the probe from sample. Oxygen molecules cross the membrane and generate a small current. Electronic circuits in the analyzer bridge detect and amplify this current. Oxygen analyzer output indications are scaled as percent oxygen. Oxygen analyzer malfunctions include con-

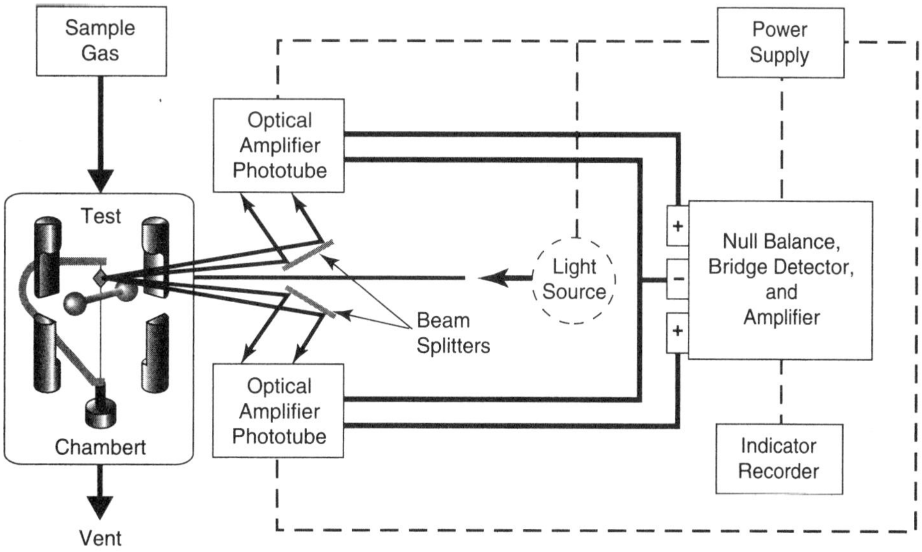

Figure 4.2-3 *Functional Block Diagram of a Paramagnetic O_2 Analyzer*

tamination and leaks into the sample line. Figure 4.2-4 shows an oxygen probe with an automatic cleaning cycle.

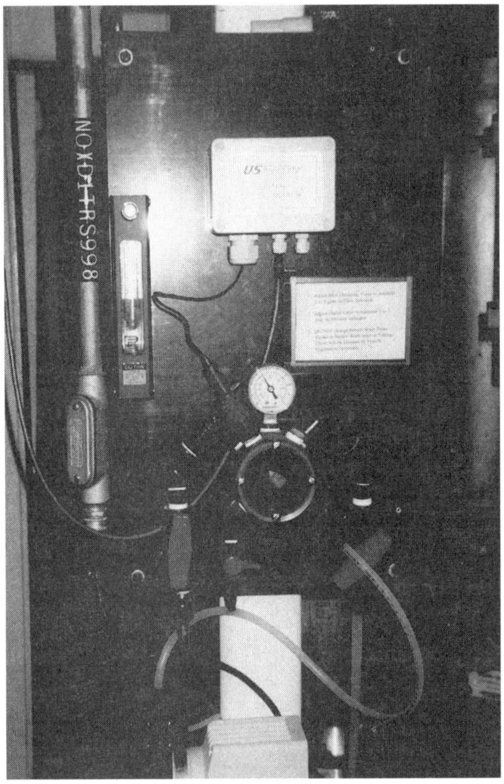

Figure 4.2-4 *O_2 Probe with Auto Clean Cycle*

4.3 Conductivity Analyzers

Conductivity analyzers detect the ability of a sample to support electrical current. Conductivity analyzers are commonly used to detect ionized particles in pure water. A pair of gold or platinum elements are precisely positioned a fixed distance apart. Sample solution is routed between the elements. Oscillating voltage is applied to the elements. Current flow between the elements is representative of the number of ions available to carry current (conductance). Conductance (S) is the reciprocal of resistance. Therefore S = 1 ÷ Ohm. The international unit of conductivity is the seimen, although it is often referred to as a mho.

Example: Absolute pure water has a specific conductance of 0.055 micro mho at 25 degrees C. To convert conductance into resistance divide the conductance into one. 1 ÷ 0.055 micro mho = 18.18 K ohms.

Figure 4.3-1 *Conductivity Sample System*

If one cubic centimeter of pure water is measured for resistance with a volt/ohm meter and the meter probes are held exactly one centimeter apart and in perfect alignment the resistance will read 18.18 K ohms. If the water is exactly 25 degrees C, the temperature compensator will read 8.55 K ohms. Extremely pure samples react more dramatically to ionic impurity than marginally contaminated samples react to additional impurity. It is also beneficial to determine the specific conductance of a sample. Specific conductance (K) is the conductance detected by one square centimeter probes through one cubic centimeter of sample. Conductivity cells are available with different degrees of sensitivity. Cell sensitivity is expressed as the cell constant. The cell constant is referenced to a standard solution of potassium chloride, and constant values are typically 1.0, 0.1, and 0.01. Analyzers are capable of using probes of any cell constant factor, but the cell constant value must be identified. Software programming or dipswitches inside the analyzer facilitate cell constant identification. Output indication values are expressed as micro mho or micro-seimen. Figure 4.3-1 shows a conductivity sampling system. Four heat exchangers at the bottom of the panel stabilize the sample temperature. Eight valves in the center of the photo regulate sample flow rate. Four resin packed columns filter and stabilize the sample. The four analyzers monitor conductivity.

4.4 Analyzer/Probe Maintenance

Modern probe analyzers are extremely accurate, but they do require proper maintenance. Erratic readings occur when sample flow rates or sample temperatures fluctuate excessively. Precise (repeatable) values are best obtained with stable sample flow rates and sample temperatures. Representative values are obtained when the probe mounting location and sample line installation provide a thoroughly mixed and contaminant free sample solution. Probes should be clean and free of foreign debris. Manufactures' specification literature provides recommended sample flow rates and sample temperatures, and also specifics proper probe cleaning techniques. The most reliable test for a probe analyzer is to compare the readings to a known good sample (standard). Many analyzers have a face mounted adjustment to correct the reading to match the standard solution value, this adjustment is called standardization and is sufficient for most processes. A more accurate method is to obtain two different standard solutions, preferably one of a low value and the other of a high value. Rinse the probe with a neutral solution and place it in the low value standard. When the reading stabilizes adjust the reference circuit to match the standard. Then, rinse the probe with a neutral solution and place it in the high value standard. Once the reading stabilizes, adjust the slope circuit to match the high value standard. If the sample delivery is within specification, the probes are clean, and the standardization procedure does not yield acceptable results, the probes or temperature compensator may be degraded. Prior to taking measurements, disconnect the probe leads from the analyzer. Temperature compensation resistance is measured with a digital voltmeter in the ohms position. The ohm value obtained is referenced to an RTD response chart and compared to the known sample temperature. Cell resistance is measured and numerically inverted into conductance. Compare the measured conductance to the known sample conductance. If the values agree, the probe is functional. To test the analyzer circuits connect a precision resistance device adjusted for the correct ohm value to the analyzer and observe the analyzer reading. If the readings agree, the analyzer is functional, and the probes are suspect. If the analyzer and probe test good, the problem is sample delivery, probe placement, or electrical noise. Probes must be kept moist at all times. Do not use a replacement probe that has dried out.

4.5 Gas Chromatography

Mikhail Tswett first developed column chromatography as early as 1906. He was able to separate plant pigments by pouring an ether solution containing leaf material into a vertical glass column packed with calcium carbonate. As the solution migrated down through the column, different pigments stratified at different heights in the column. This resulted in a chromatogram, a series of horizontal bands of color along the height of the column with each band composed of a different pigment.

One of the most valuable analyzers in the process industry is the gas chromatograph. While most analyzers are instruments that directly monitor the product and provide instantaneous analysis, chromatography is actually an analytical process that samples, separates, detects, qualifies, and quantitates a product. The chromatograph consists of a programming controller section, a sample collection section, a separation section, and a detection section. Refer to block diagram Figure 4.5-1.

The program control section provides the logic sequence to automatically manipulate the sample and column valves. It contains the timer, the power supply for the bridge circuits, the automatic zero mechanism, adjustment pots, and signal conditioning circuits. Several

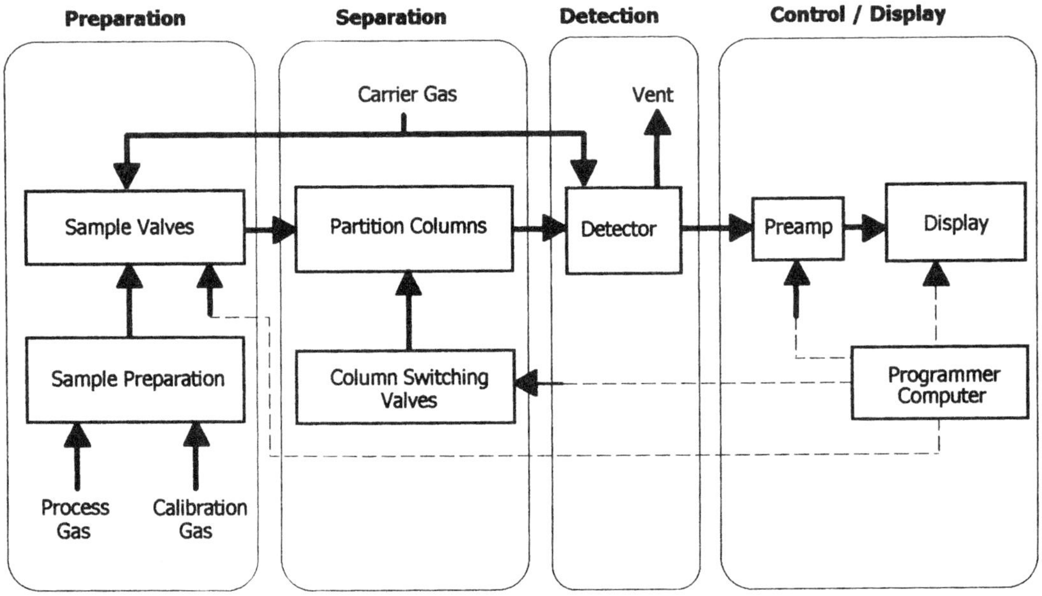

Figure 4.5-1 *Gas Chromatographic Analyzer Block Diagram*

automatic functions are switch selectable. Typical programs include; chromatogram mode, calibration mode, or automatic analysis mode. Programmable temperature control and automatic stream switching is also supported. The output display device is usually a strip chart recorder.

The sample collection section consists of the piping, tubing, valves, filters, dryers, and heaters needed to collect, prepare, and inject the sample and other gases. Product samples are syringe injected directly into a septum on laboratory analyzers. On-line analyzers use sample valves to introduce a fixed volume sample into a carrier gas stream that transports the sample through the column. Multi-position valves alternately route purge gas, calibration gas, and sample gas through the column. The samples are filtered, pressure and temperature stabilized, and forced through the column by an inert carrier gas or liquid mobile phase. Typical carrier or mobile phase gasses are nitrogen, helium, and argon. Carrier gasses are application specific and must not react with any component of the sample. They must be 99.99% pure and completely moisture free.

The separation section consists of various columns that segregate discreet components from the sample effluent. The sample is separated at this mobile phase interface by adsorption. Adsorption is the inclination of some substances to adhere to the surface of a particular media. This effect results in a stratification of compounds traversing a column of an adsorbent media. Compounds with no adsorbent reaction to the obstructive media travel through the column unimpeded. Compounds that are sensitive to the adsorbent media resist propagation and stratify along the column in degreed order of adsorption. Columns are made of small diameter stainless steel tubing (1/8 inch or 1/4 inch) packed with different media to segregate particular substances, and wound into coils to conserve space, see Figure 4.5-2. Examples of stationary packing are porous silica microspheres, activated alumina, and activated carbon. Porous silica microspheres are often chemically coated with very selective compounds (substrates) to modify their characteristics. Continued introduction of additional reagent gas or solvent to the column forces the segregated components to continue their

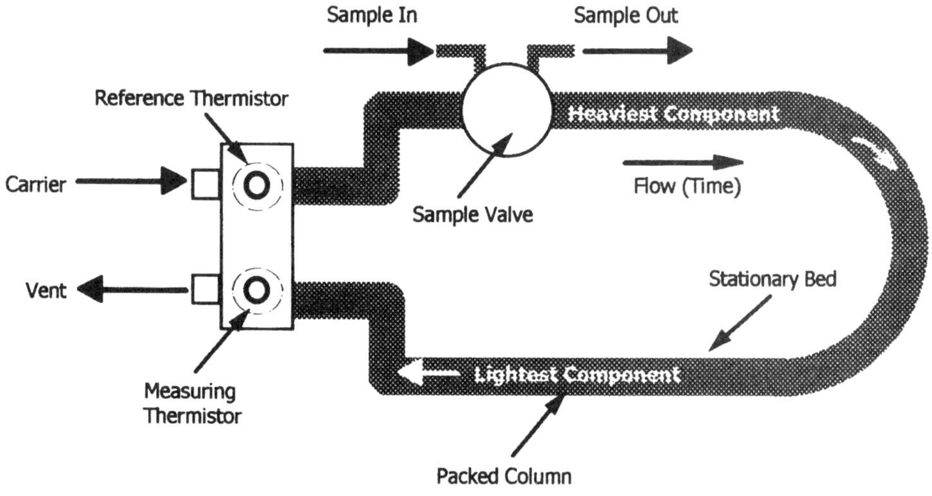

Figure 4.5-2 *Simplified Gas Chromatograph Sample Column*

migration through the packing. The sweeping of components from the column is called elution. The time from sample injection to the detection of a component is the elution time. Sample solution exiting the column contains the separated (eluted) compounds in relation to their propagation time. Elution time identifies the molecular size of the component. Components with the lightest molecular weight transverse the column first. Successive components eluted from the sample exit the column in order of increasing molecular weight. Elution time and component separation are a function of the selected packing and column length. If only lighter components are to be analyzed, a precut (or stripper) column is placed ahead of the analysis column to separate the lighter, faster components from the heavier slower components. As soon as the light components exit the stripper column and enter the analysis column, column valves reverse flow to the stripper column and purge the heavy components from the column. Once purged of heavy components, the stripper column is placed back in service. By not having to wait on the heavy components to completely elute the entire column, sampling time is reduced and sampling rates may be increased, thus resulting in increased efficiency.

4.6 Chromatograph Detectors

As the carrier/elute mixture exits the column it is routed through the detection section. A detector measures each component in its separated state. The most common chromatograph detectors in use today are hydrogen flame ionization detectors, spectrometers, and thermal conductivity detectors. Flame ionization detectors are capable of detecting component concentrations as low as 0–1 ppm. As depicted in Figure 4.6-1, the detector consists of a chamber with inlets for hydrogen, column effluent (sample), and air, and an outlet vent. Inside the chamber is an air diffuser, a burner, an igniter, and the detection electrodes (anode/cath-

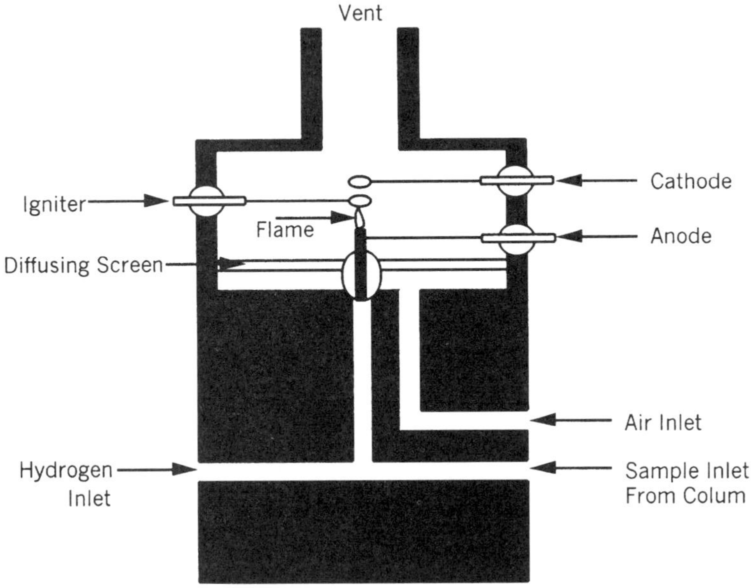

Figure 4.6-1 *Hydrogen Flame Ionization Detector*

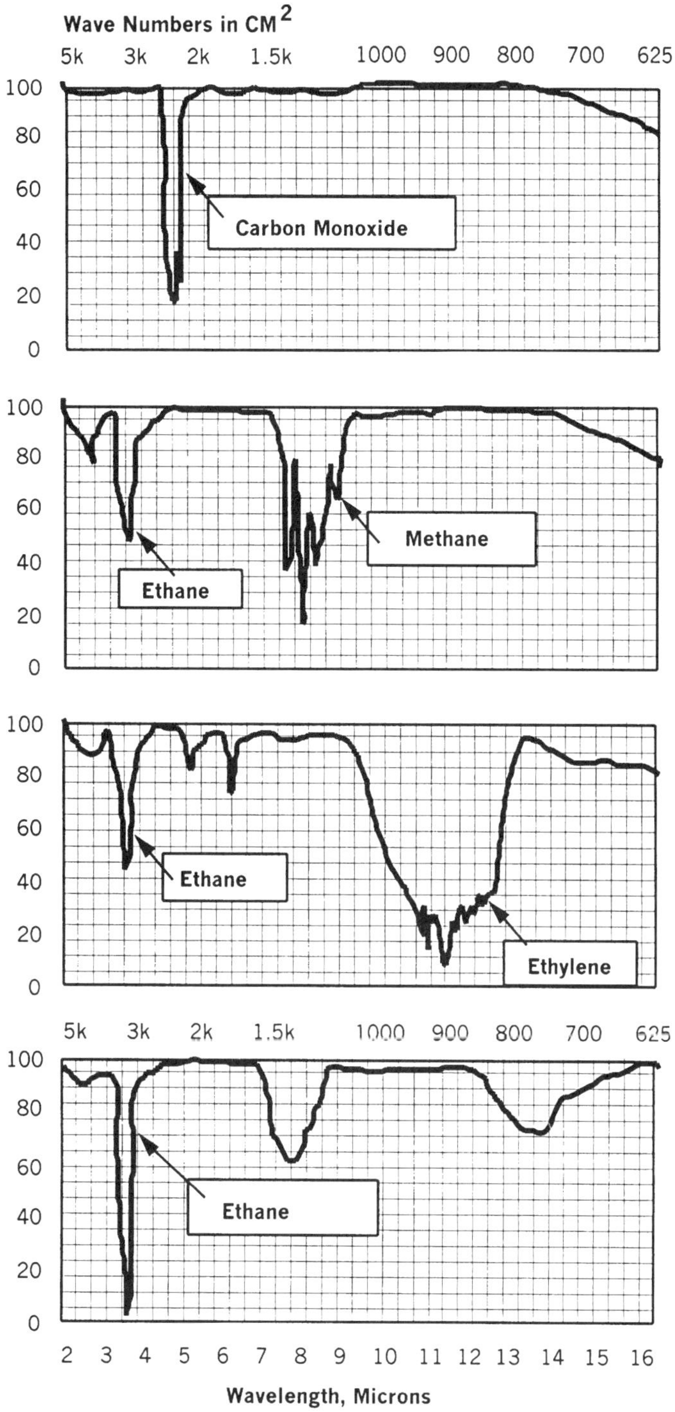

Figure 4.6-2 *Typical Infrared Spectra*

ode). A DC voltage (100 –1500 volts) is applied to the electrodes. The hydrogen flame induces the emission of electrons from the hydrocarbon molecules in the effluent stream. The ions are collected on the electrodes and amplified in a high quality electrometer amplifier. The sensitivity of the detector is proportional to the carbon content of the mixture.

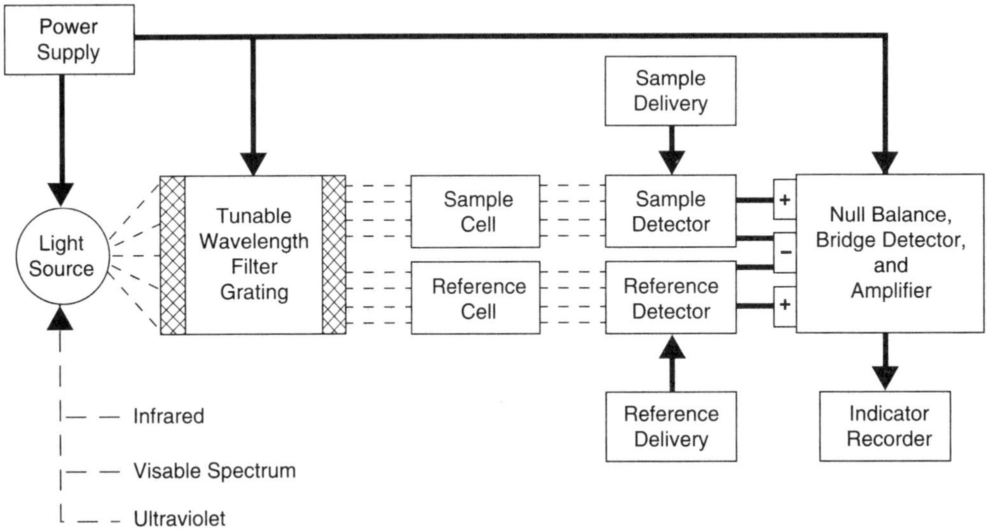

Figure 4.6-3 *Functional Block Diagram of Spectrometer*

Infrared analyzers (spectrometers) are used to detect components that exhibit absorptive characteristics in the infrared region of the wavelength spectrum. When a molecule of gas (or liquid) is subjected to electromagnetic radiation it will selectively and disproportionably absorb a single specific wavelength of energy. Figure 4.6-2 compares the absorptive characteristics of several gases. This phenomenon permits qualitative analysis by spectral reactivity. The inverse of absorbance is transmittance. Transmittance is the fraction of radiant energy that, having entered a layer of absorbing matter, reaches its farther destination. Since the degree of transmittance is relative to sample concentration, this process also yields quantitative data. Plotting transmittance against wavelength results in a spectrogram. Spectral analysis may be performed with a variety of wavelengths including ultra-violet, visible light, and infrared. Elemental gases, hydrogen, nitrogen, oxygen, chlorine and the rare gases do not absorb energy in the infrared region. The spectrometer consists of a light source, a tunable grating or filter, a sample flow cell with detector, and a reference flow cell with detector, see Figure 4.6-3. The source emits a spectrum of radiant energy (light). Tunable gratings and filters isolate the precise wavelength absorbed by the product detected. This frequency is then transmitted simultaneously through the sample cell and the reference cell. The difference in exit energy detected between the sample cell and the reference cell is proportional to the energy absorbed by the sample. Since a linear relationship exists between absorption and concentration, quantitative analysis is permitted.

Thermal conductivity analyzers are also used to detect components of chromatograph effluent streams. Thermal conductivity detectors consist of a Wheatstone bridge with a thermistor mounted in the measuring junction. The bridge is balanced to a null condition with pure carrier gas passing over the measuring thermistor. Sample component gases are less thermally conductivity than carrier gases. As the component gas passes over the measuring thermistor, the thermistor's temperature rises decreasing the thermistor's resistance. The

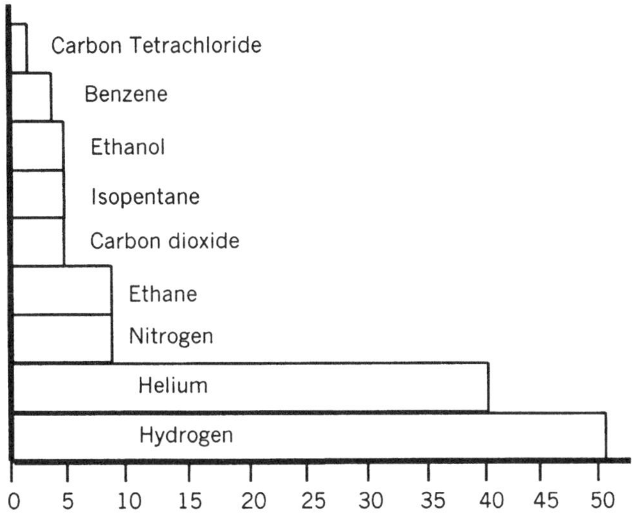

Figure 4.6-4 *Relative Thermal Conductivity of Gases*

resulting bridge unbalance is representative of the sample concentration and elution time is representative of the component identity. Supporting amplifiers processes the signal and the results are printed to a chart or bar graph.

Gas chromatograph malfunctions are usually related to sample delivery problems, sample contamination, or column packing contamination. Calibrations are performed by flowing special calibration gasses through the analyzer and adjusting for zero and span. Significant elements of chromatographic analysis are time, temperature, pressure, and flow. Practical applications strictly regulate sample temperature, pressure, and flow values. Sample propagation time therefore, is the discriminating variable.

4.7 The Distillation Process

The distillation process adds heat energy to a liquid mixture forcing the compounds within the mixture to vaporize in order of volatility. The separated compounds are then condensed back into their liquid state. Volatile compounds readily transition to their gaseous state, while the heavier compounds tend to remain in the mixture. A popular distillation process increases the alcohol content of spirits by heating the water and alcohol mixture sufficiently to gas off the alcohol. The alcohol is then condensed (cooled) back into a liquid. Since the boiling points of water and alcohol are relatively close, the distillation process always collects some water and/or leaves behind some alcohol. Several repeat distillations are required to obtain significantly high alcohol concentrations. Simple distillation is generally sufficient to separate a single compound from a mixture.

Several different compounds are simultaneously extracted from a mixture by the fractional distillation process. Fractional distillation applies heat energy to a mixture just as simple

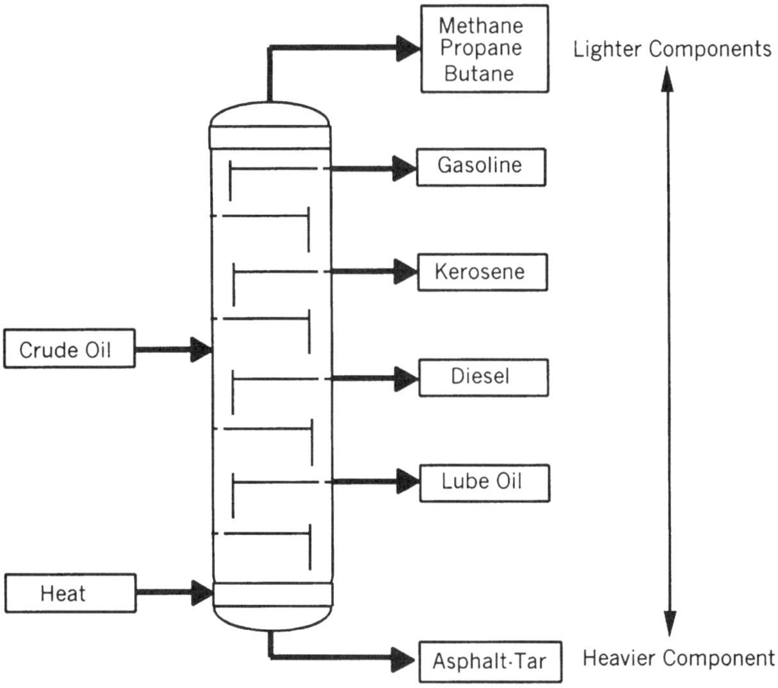

Figure 4.7-1 *Fractional Distillation Column*

distillation. Extracted vapors are then routed through a packed vertical column. Separated vapors stratify and congregate at specific levels in the column. Lighter compounds migrate to the top of the column while heavier compounds remain near the bottom. Extraction taps are located at different levels along the column. Each separated vapor is then drawn off and condensed into a pure product. Figure 4.7-1 depicts a simplified diagram of a fractional distillation column. Fractional distillation is an efficient process and is commonly used to refine petroleum. Gas chromatographic analysis can determine if a sample has components of ethane, ethylene, and propylene and the concentration of each component. This trace detection ability is used to determine stratification of substances within a fractioning column. The amount of heat energy applied to the mixture determines the level of stratification within the fractioning column. If less heat is applied, the substances stratify at a lower level within the fractioning column. If more heat is applied the substances stratify at a higher level within the fractioning column. Gas chromatograph sample lines are installed at each extraction tap. The chromatographs are configured to detect the predominate substance of the next lower or higher stratification level. A chromatograph configured to detect the substance at the next lower level of stratification expects to detect a minute trace of that product. If no trace is detected, the temperature is to low. If a significant amount is detected, the temperature is too high. A heat control signal is generated to correct temperature and bring the stratifications within the column back to the appropriate level.

Chapter Four Summary:

- Chemical analysis is the science and technology used to identify and measure the chemical composition of a product.
- Qualitative analysis determines the presence of a particular substance in a sample. Quantitative analysis determines the amount of a particular substance in a sample.
- Common industrial applications of chemical analysis include monitoring for product purity, verifying concentration of solutions, and detection of unwanted or undesirable contaminates.
- Unstable compounds can exert tremendous energy if allowed to oxidize, and combustible processes can spontaneously ignite if exposed to oxygen.
- Measurement cells detect the concentration or activity of the sample, reference cells generate a constant fixed reference voltage, and temperature compensators provide a temperature feed-back signal.
- Most reactions are temperature sensitive and require accurate temperature sensing feed-back.
- Bridge circuits located in the analyzer detect the minute changes in cell signal values.
- pH analyzers measure the alkalinity or acidity of an aqueous solution.
- Highly acid concentrations are referenced at (0) on the pH scale. Pure water is neutral (7), and highly alkaline (base) concentrations are referenced at (14).
- An oxygen permeable membrane isolates the oxygen probe from the sample. Oxygen molecules cross the membrane and generate a small current.
- Conductivity analyzers detect the ability of a sample to support electrical current. Conductivity analyzers are commonly used to detect ionized particles in pure water.
- Conductivity (S) is the reciprocal of resistance. Therefore $S = 1/\text{Ohm}$. The international unit of conductivity is the seimen, although it is often referred to as a mho.
- Precise (repeatable) values are best obtained with stable sample flow rates and sample temperatures.
- Representative values are obtained when the probe mounting location and sample line installation provide a thoroughly mixed and contaminant free sample solution.
- Probes should be clean and free of foreign debris.
- The most reliable test for a probe analyzer is to compare the readings to a known good sample (standard).
- A face mounted adjustment to correct the reading to match the standard solution value is called standardization.
- The spectrometer consists of a light source, a tunable grating or filter, a sample flow cell with detector, and a reference flow cell with detector.
- The difference in exit energy detected between the sample cell and the reference cell is proportional to the energy absorbed by the sample.

- Charts are plotted that reference the amount of absorbance versus travel time. Transition time identifies the product, and absorbance delineates the concentration.
- Adsorption is the inclination of some substances to adhere to the surface of a particular media.
- The stratification of compounds traversing a column of an adsorbent media results in a chromatogram.
- Chromatography is an analytical process that samples, separates, detects, qualifies, and quantitates a product.
- Column valves alternately route purge gas, sample solution, and sample gas through the column.
- Samples are forced through the column by an inert carrier gas or liquid mobile phase.
- Typical carrier or mobile phase gasses are nitrogen, helium, and argon.
- A column of stationary phase packing separates the components in the sample.
- Columns are packed with different media to segregate particular substances. Examples of stationary packing are porous silica microspheres, activated alumina, and activated carbon. Porous silica microspheres are often chemically coated with very selective compounds (substrates) to modify their characteristics.
- Sample solution exiting the column contains the separated compounds in relation to their propagation time.
- Gas chromatograph malfunctions are usually related to sample delivery problems, sample contamination, or column packing contamination.
- Significant elements of chromatographic analysis are time, temperature, pressure, and flow. Practical applications strictly regulate sample temperature, pressure, and flow values. Sample propagation time therefore, is the discriminating variable.
- The distillation process adds heat energy to a liquid mixture forcing the compounds within the mixture to vaporize in order of volatility. The separated compounds are then condensed back into their liquid state.
- Several different compounds are simultaneously extracted from a mixture by the fractional distillation process.
- Chromatographs sample for trace elements of petroleum distillates at each level of a fractioning column and develop a temperature control demand signal to maintain proper stratification within the column.
- Chromatographs are configured to detect the predominate substance of the next lower or higher stratification level. A heat control signal is generated to correct temperature and bring the stratifications within the column back to the appropriate level.

Chapter Four Review Questions:

1. State the definition of chemical analysis.
2. State the purpose of qualitative analysis.
3. State the purpose of quantitative analysis.
4. List several industrial applications of chemical analysis.
5. State the purpose of analytical instruments.
6. State the undesirable effect of introducing oxygen into some processes.
7. List three types of detectors that comprise common analytical probe assemblies.
8. State the purpose of a measurement cell.
9. State the purpose of a reference cell.
10. State the purpose of a temperature compensation detector.
11. Explain the relationship between potential difference and solution concentration.
12. State the purpose of pH analyzers.
13. Identify the acidity or alkalinity of a substance based on it's pH value.
14. State the purpose of oxygen analyzers.
15. State the purpose of an oxygen permeable membrane in an oxygen probe.
16. State two common oxygen analyzer malfunctions.
17. State the purpose of conductivity analyzers.
18. Describe the relationship between conductivity (S) and resistance.
19. State two terms used to indicate conductivity.
20. State the definition of specific conductance (K).
21. List several sample criteria required to produce precise analytical values.
22. State the most reliable method to determine operability of an analyzer probe.
23. State the adjustment used to correlate an analyzer reading to a known sample.
24. Describe the troubleshooting technique used to isolate a reference temperature malfunction.
25. Describe the distillation process.
26. Describe the fractional distillation process.
27. Differentiate between simple distillation and fractional distillation.
28. Describe the purpose of chromatographs in the fractioning process.

29. Explain trace detection as used by chromatographs in control applications.
30. Define adsorption.
31. Explain the application of adsorption as used in chromatographic analysis.
32. Define chromatogram as used in chromatography.
33. State the purpose of chromatography as used in industrial applications.
34. List three steps of the chromatographic process.
35. State the purpose of a sampler valve.
36. State the purpose of the mobile phase.
37. List several carrier gasses.
38. State the purpose of the stationary packing.
39. List several examples of stationary column packing.
40. Describe the relationship between absorption and sample transmission time.
41. State the purpose of the spectrometer as it applies to chromatography.
42. List four functional sections of a spectrometer.
43. Describe the function of a spectrometer.
44. Explain the relationship between travel time and concentration.
45. List several potential chromatograph malfunctions.
46. List the significant sample parameters controlled in chromatographic analysis.
47. State the primary discriminating sample variable in chromatographic analysis.

Chapter 5

Controllers

OBJECTIVES

Upon completion of Chapter Five the student will be able to:

- *Describe the basic functions of pneumatic, electronic, and digital controllers.*
- *List the operational modes of controllers.*
- *Describe automatic control mode.*
- *Describe manual control mode.*
- *Describe cascade control mode.*
- *Develop the proportion bands (1/Gain) P response to a step change in input.*
- *Develop the proportion bands (1/Gain) P response to a ramp change in input.*
- *Develop the integral (Reset) I response to a step change in input.*
- *Describe reset wind-up.*
- *State the effects of reset wind-up on controller output.*
- *Develop the integral (Reset) I response to a ramp change in input.*
- *Develop the derivative (Rate) D response to a step change in input.*
- *Develop the derivative (Rate) D response to a ramp change in input.*
- *State the difference between controller output and actual control valve position in the field.*

- *List the functional components of a pneumatic controller.*
- *List the control and tuning adjustments available on a pneumatic controller.*
- *State the location of control and tuning adjustments on a pneumatic controller.*
- *Describe the function of pneumatic controller components including the input bellows, flapper nozzle, force balance beam, feed-back bellows, and output relay.*
- *List the functional components of an electronic controller.*
- *List the control and tuning adjustments available on an electronic controller.*
- *State the location of control and tuning adjustments on an electronic controller.*
- *Describe the function of electronic controller components including the voltage divider set-point, error detection amplifier, deviation circuit, integral circuit, and output driver.*
- *List the functional components of a digital controller.*
- *List the control and tuning adjustments available on a digital controller.*
- *State the location of control and tuning adjustments on a digital controller.*
- *Describe the function of digital controller components including the analog to digital converter, memory set-point, binary counter, algorithm processor, and digital to analog converter.*
- *List the functional components of a local controller.*
- *List the control and tuning adjustments available on a local controller.*
- *State the location of control and tuning adjustments on a local controller.*
- *Describe the function of local controller components.*
- *List the common malfunctions associated with pneumatic controllers.*
- *List the common malfunctions associated with electronic controllers.*
- *List the common malfunctions associated with digital controllers.*

5.0 Introduction

Controllers provide process variable monitoring (input signal), set-point to process comparison (error signal), and developed correction signal algorithms (output). Transmitters and transducers provide the input signal to the controller. The controller receives and accepts the input as an accurate representation of the actual process variable monitored. The controller then compares the input signal to a predetermined optimal or desired value, the set-point. If a difference between set-point and input is detected, the controller computes and applies a specific preset algorithm output to eliminate the difference. The correction signal is transmitted to a final control element that manipulates the controlled variable to effect the correction. Controllers monitor inputs continuously, and automatically respond to any changes. Specific correction factors required for each process vary. Several error response actions act individually, or combine, and provide the optimum correction signal. Scaled dials mounted directly on the controller provide response tuning and adjustment capability. Intelligent controllers store these values in memory and require tuning through a menu-driven keypad or a communicator device. Position switches select automatic or manual control mode operation. Internal tracking of the output signal while in manual mode ensures a bumpless transfer when the controller is restored to automatic operation. bumpless transfer is the ability to switch between automatic and manual control operation without disturbing the process or repositioning the final control element.

5.1 Controller Response

Controllers operate in several selectable modes. The most common are automatic, manual, and cascade. Automatic control is generally preferred. In this mode the controller performs continuous monitoring and manipulation of the process as required to maintain set-point. Manual control overrides the automatic functions of the controller and allows the operator to use the controller output adjustment to position the final element in the desired position. The automatic section of the controller tracks the manual output, and monitoring and recording functions are maintained. Cascade control is similar to automatic control, except the local set-point is disabled and another controller remotely controls the set-point.

Proportion band (1/gain × 100) P (sometimes represented as PB), integral (Reset) I, and derivative (Rate) D are the three most common response characteristics. Proportional values are expressed as percent of scale range, and are generally between 0% and 500%. Proportional response characteristics are often expressed as gain. The reciprocals of these terms refer to identical response values. The terms are interchangeable providing correct conversions are applied. Proportion band is equal to 1/gain × 100 and gain is equal to 1/p × 100 band. Proportion band determines response slope relationship between error, in percent of scale, and output, in percent of output range. Proportional correction is applied instantaneously, and the response slope is fixed. A proportion band of 100% responds to a 100% error with a 100% change in output. A proportion band of 50% responds to a 50% error with a

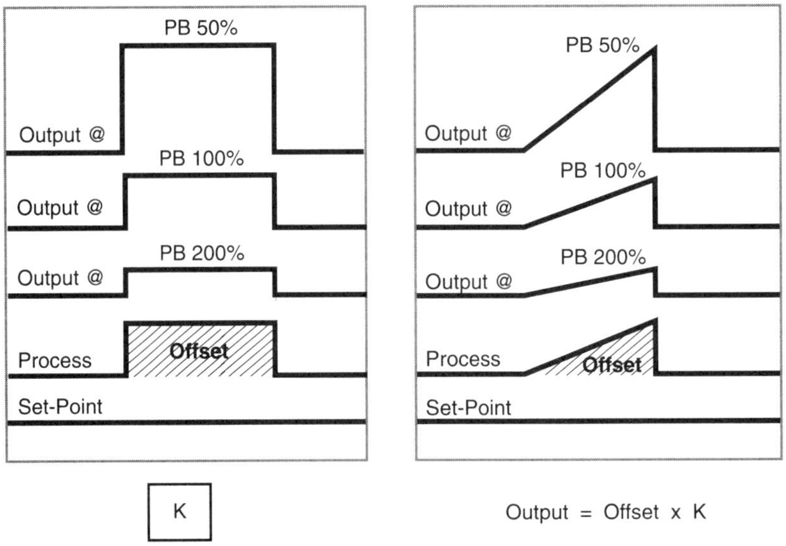

Figure 5.1-1 *Proportional Response*

100% change in output. A proportion band of 200% responds to a 200% error with a 100% change in output, Figure 5.1-1.

Integral (reset) is a time-weighted response. Since proportional response is instantaneous and fixed, proportional action alone may successfully stabilize the process at a value other than set-point. Integral action develops a correction value determined by the amount of error

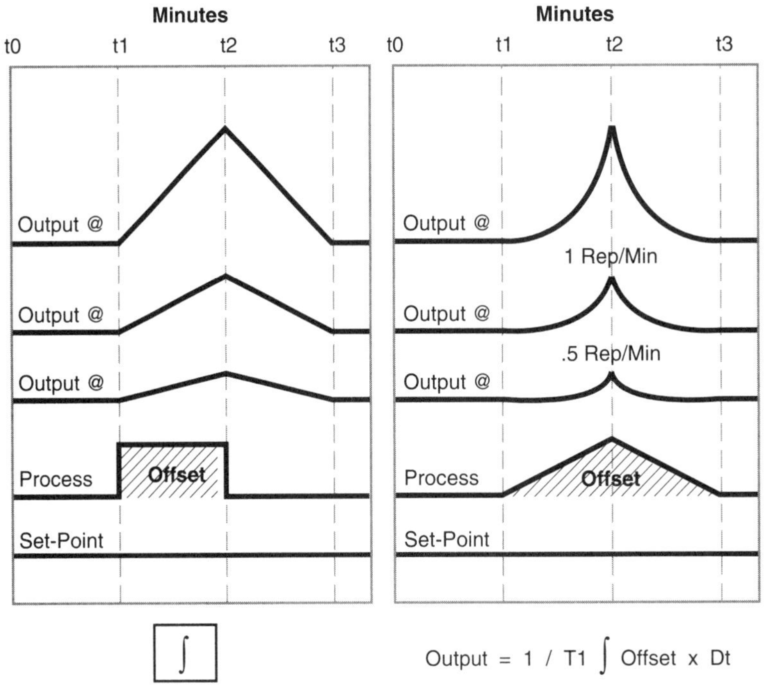

Figure 5.1-2 *Integral Response*

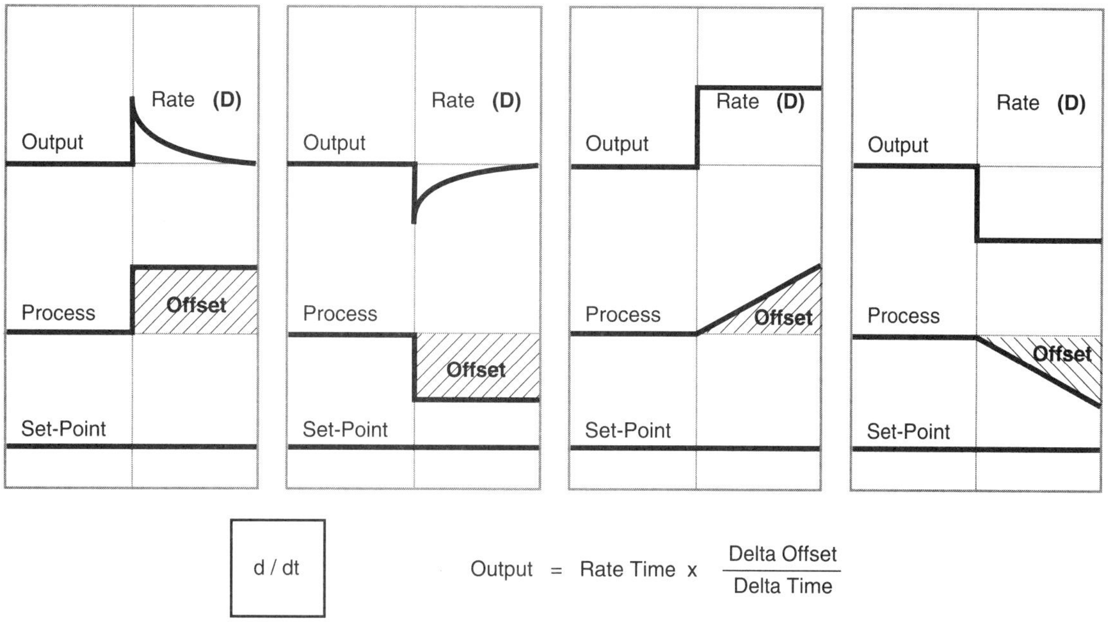

Figure 5.1-3 *Derivative Response*

over time. Integral values are expressed as minutes per repeat or repeats per minute. The term repeat refers to the error signal. Integral action restores process to set-point by complimenting the original input signal value with a ramped amount of the error value and driving the output higher. A setting of one repeat per minute adds one error value to the output each minute, Figure 5.1-2. The error is reduced through each correction cycle (repeat) until offset is eliminated. Reset output continues to increase over time until set-point deviation is eliminated. If the final control element fails to correct the offset, reset output saturates. This phenomenon is called reset wind-up. A controller in a reset wind-up condition does not respond to additional process changes until sufficient reset time integrals eliminate the wind-up.

Derivative (rate) is also a time-weighted response. Derivative action develops correction values relative to the rate of change of the input value, Figure 5.1-3. Derivative response is developed independently of error magnitude. This response characteristic is often referred to as lead. Proportional and integral are reactive correction characteristics, both actions only initiate in response to measured error. Derivative is a pre-active correction characteristic; response is initiated by a change of input value. Rapid input changes initiate an aggressive derivative response, and moderate to gradual input changes initiate minimal to no derivative response.

The magnitude of each individual response characteristic is summed into a final cumulative value that constitutes the control algorithm, Figure 5.1-4. Controllers apply the algorithm to maintain process-input values at set-point. A correction signal is developed in response to deviation and the effectiveness of correction is determined by continually comparing process input (feed-back) to set-point. Adjustments to the PID settings are referred to as tuning the controller.

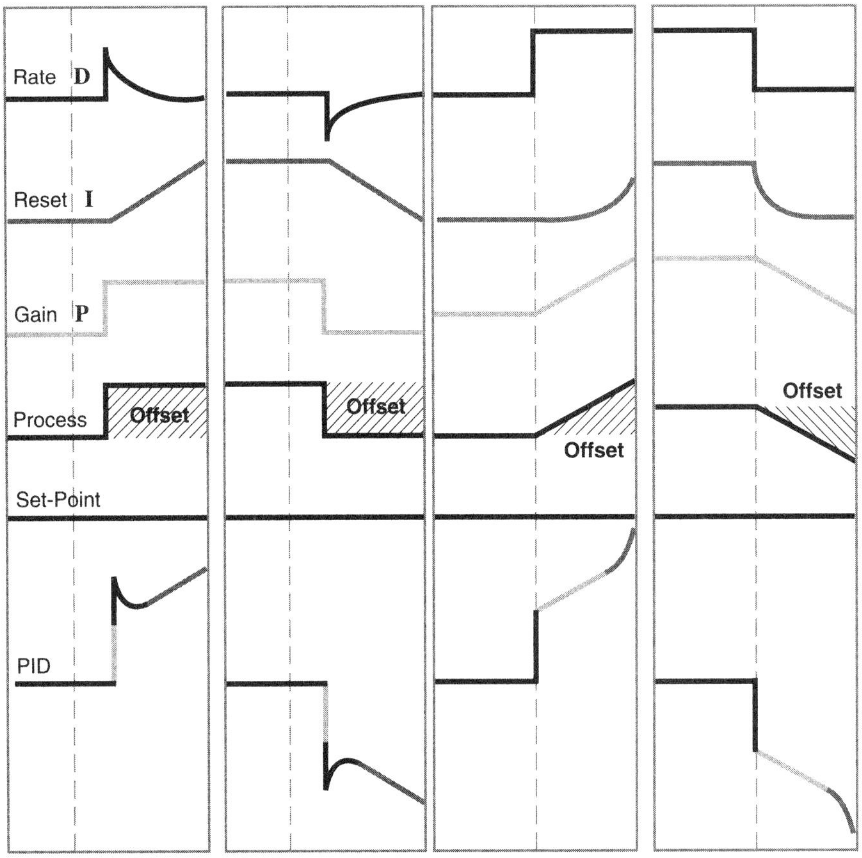

Figure 5.1-4 *Combined Response*

The controller effects all corrective action through a final control element. Ultimately loop stability and control is realized through final control element characteristics. Due to the variety of tuning parameters applied to different processes, control valve position is not always directly equivalent to controller output. The control valve positioner characterizes control valve position. The positioner accepts the demand (correction) signal from the controller and positions the valve according to a designed response profile. Common profiles are linear, quick opening and equal percentage. These profiles are investigated further as they apply to final control elements.

5.2 Functional Components of Controllers

Pneumatic controllers monitor and compare a process variable to a set-point, detect the error, and apply an output algorithm to eliminate the error. Pneumatic controllers display the process variable value, set-point, and analog output value. Set-point adjustment and auto/manual transfer are also face mounted, Figure 5.2-1. Many controllers contain chart drive mechanisms to record process fluctuations. The chart drive is a clock driven gear reduction roller that feeds scaled graph paper under the ink fed controller-indicating pens.

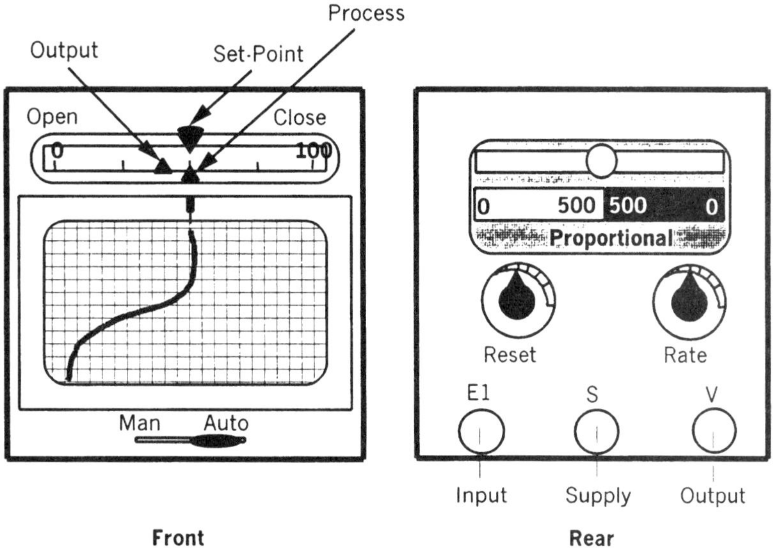

Figure 5.2-1 *Pneumatic Controller*

PID tuning and chart drive on/off controls are located either on the rear or inside of the case. Partial removal of the controller from the panel may be required to access these controls.

The most common pneumatic signal is 3–15 psig. Pneumatic controllers incorporate a bellows assembly as the input sensor. The bellows converts input pressure into a physical displacement or motion to the balance beam, Figure 5.2-2. The beam is attached to the bellows at one end and a flapper/nozzle at the other. The beam pivots on a leverage multiplier fulcrum. Bellows induced displacement at one end of the beam is detected as an error signal by the flapper nozzle assembly. The controller is supplied with 20 psig of continuous regulated

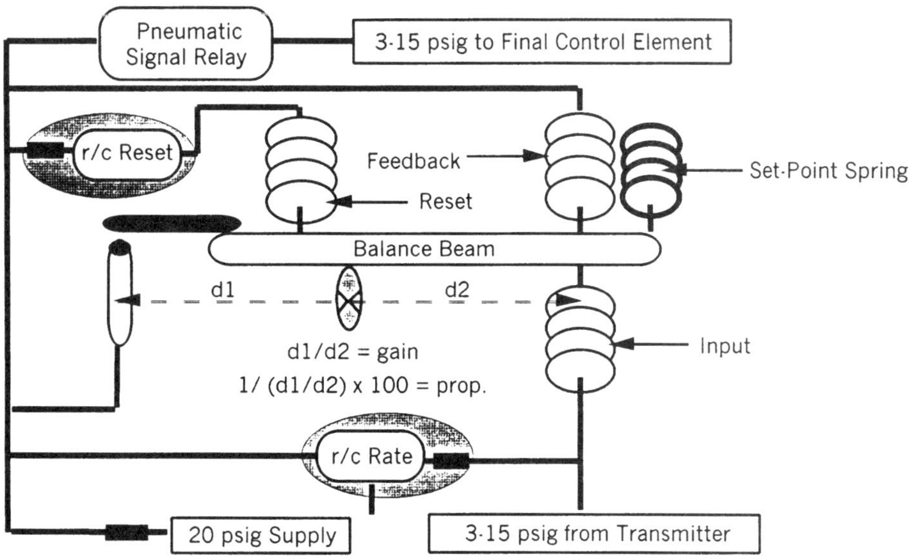

Figure 5.2-2 *Pneumatic Control Beam Assembly*

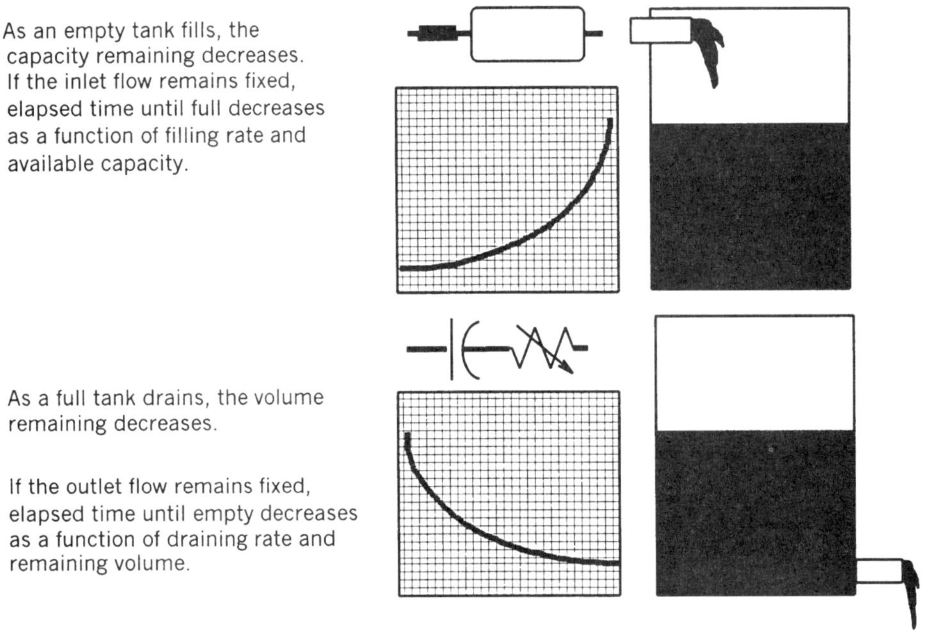

Figure 5.2-3 *Resistive/Capacitive Timing Circuits*

instrument air (supply). The measured variable value from the transmitter is applied to the input bellows (input). The bellows transfers an equivalent amount of energy to a force balance mechanism. A flapper nozzle assembly detects the amount of deviation at the force balance beam. The flapper nozzle assembly backpressure is amplified by a pneumatic relay and retransmitted as a pneumatic output signal. Supply air flows through a restriction screw and is supplied in parallel to the nozzle, feed-back bellows, and upper chamber of the output relay. The flapper acts to block the nozzle and build backpressure (error signal). The feed-back bellows and output relay simultaneously sense the error signal. The feed-back bellows expands to restore balance to the beam. The relay amplifies the error signal and outputs a representative value of supply air. Fulcrum position on the beam determines proportional band through simple lever ratios. A restrictor/tank R/C circuit, Figure 5.2-3, develops reset. The time constant of the tank network is a function of restrictor flow times tank capacity. Increasing or decreasing the amount of restriction adjusts reset action. The reset tank transmits to a reset bellows located on the beam to compliment the input bellows force. A tank circuit also develops the rate function. The rate tank, however, accepts input pressure and applies it directly to the output. Some rate tanks have an additional restrictor installed as a constant bleed.

Electronic Controllers perform the same basic functions as pneumatic controllers. Although executed by electronic circuits, the control functions of proportional, integral, and derivative are identical. The analog process variable display output-level indicator, set-point indicator, set-point adjustment, and auto-manual transfer switch is located on the front of the controller. The chart drive data logger mechanism employs a printer cartridge or a thermal writing device. A mechanical latch located on the front of the controller secures the controller

Figure 5.2-4 *Level Controller*

in the panel. Once released, the controller is partially pulled from the panel to access the tuning knobs located inside the case, Figure 5.2-5.

A differential operational amplifier circuit, Figure 5.2-6, generates proportional band. A voltage divider circuit on the reference input establishes set-point, and a feed-back resistor develops gain (proportion). An electrical R/C tank composed of a resistor and a capacitor develops integral. A R/C tank also develops derivative.

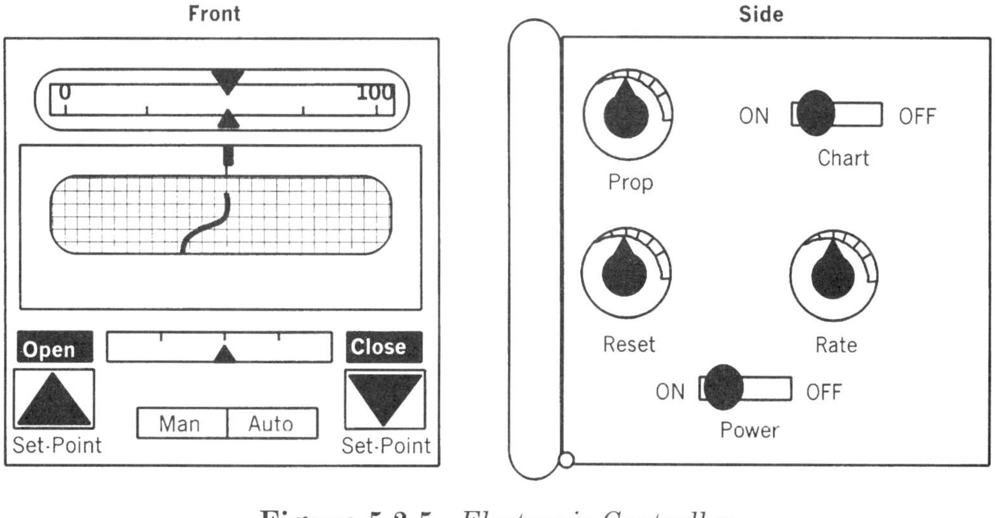

Figure 5.2-5 *Electronic Controller*

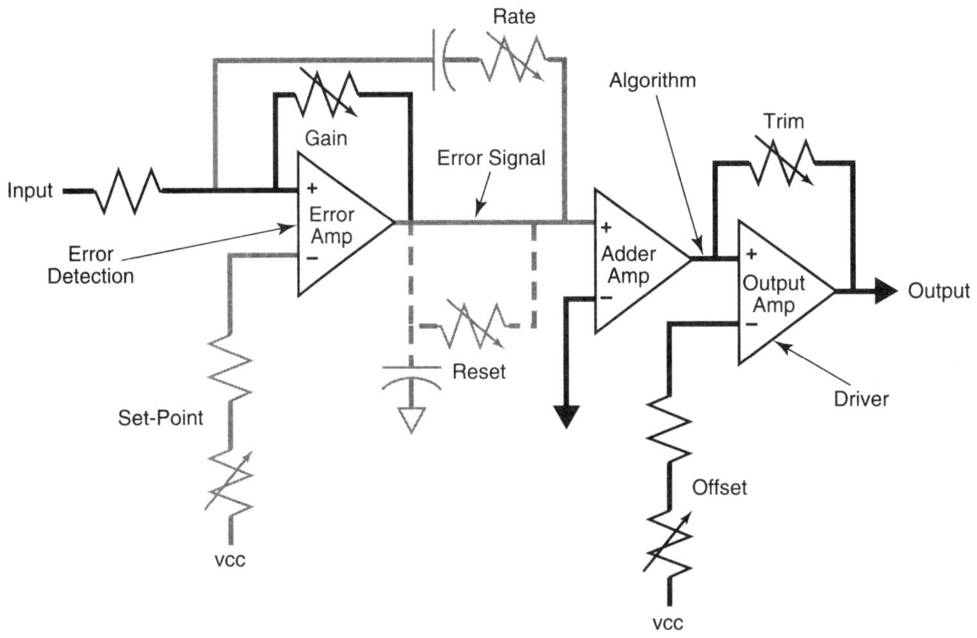

Figure 5.2-6 *Operational Amplifiers*

Digital controllers perform the same basic function as pneumatic and electronic controllers; they do however have several added advantages. Digital controllers offer the ability to configure an array of functions. Digital controllers are also capable of communicating in

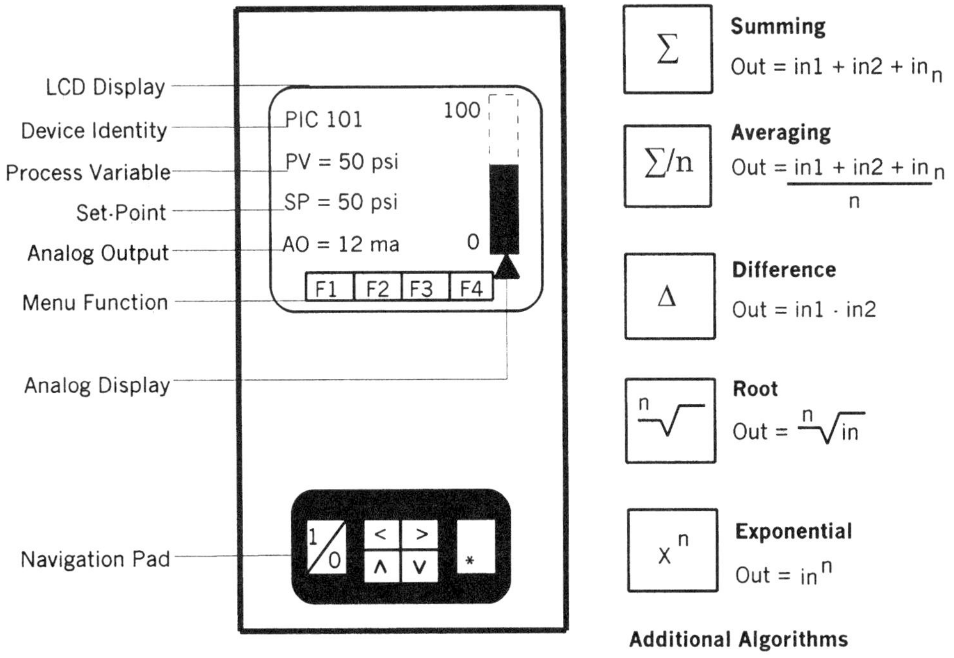

Figure 5.2-7 *Digital Controller*

multi-drop configurations, and performing auto tuning and self-diagnostics. Analog process value, set-point, and output value are displayed on a LCD screen, Figure 5.2-7. Data trends are stored in memory or written to a floppy disk. Tuning parameters are accessed through a menu driven keypad or a digital communicator. Signal processing is accomplished through the use of several standard digital components. The analog to digital converter accepts the process variable and converts the analog value into an equivalent binary count. The count is then transmitted to a binary counter/comparator that determines the difference between the process value and the memory set-point stored in a register. The algorithm processor uses the extracted error value to develop the appropriate correction value. The final digital correction signal is then converted back into an analog value by the digital to analog converter and transmitted to the final control element by an isolated output driver.

5.3 Local Controllers

Local controllers perform identical functions as board mount controllers, they are however enclosed in weather tight housings and mounted in the field near the controlled device. Local controllers provide local indication of process variable value, set-point adjustment and indication, output indication, manual output adjustment, and adjustment dials for proportional, integral, and derivative. Many local controllers provide output balance indicators to assist bumpless transfer functions, Figures 5.3-1 and 5.3-2. Recording controllers trend data on a circular chart.

Local controllers are often physically mounted, as shown in Figure 5.3-3, to the final control element (air-operated valve). The controller input is connected to the process line and the

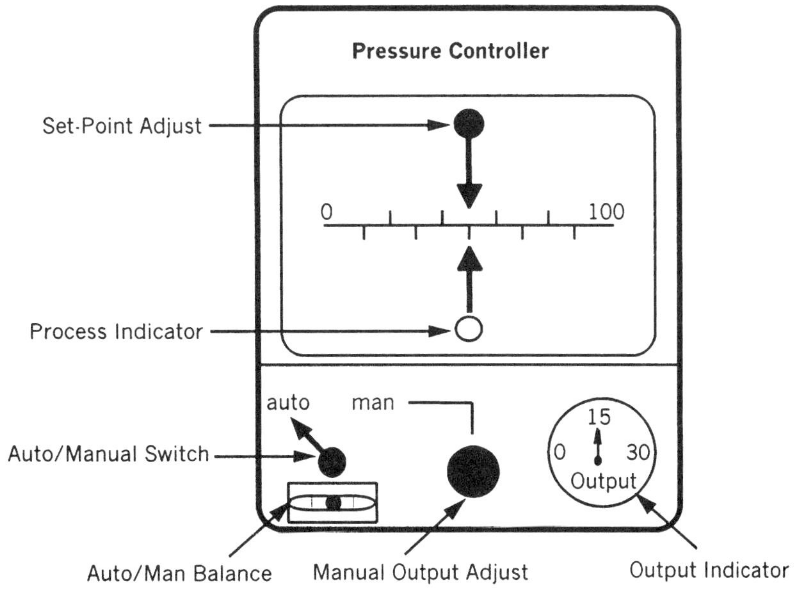

Figure 5.3-1 *Local Controller*

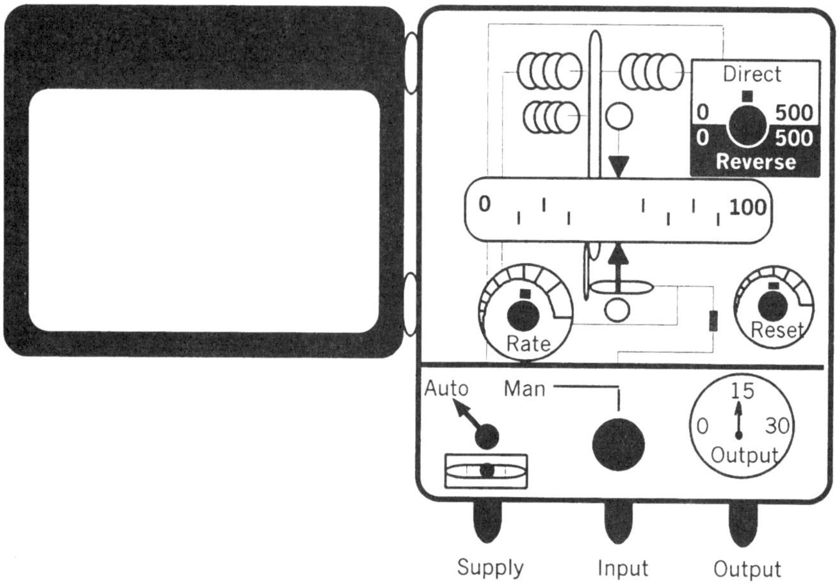

Figure 5.3-2 *Local Controller Adjustments*

process variable is detected directly at the controller. The controller output is transmitted directly to the valve actuator. The valve is adjusted to position from full open to full closed in relation to the local controller output. The need for a transmitter in the control loop is

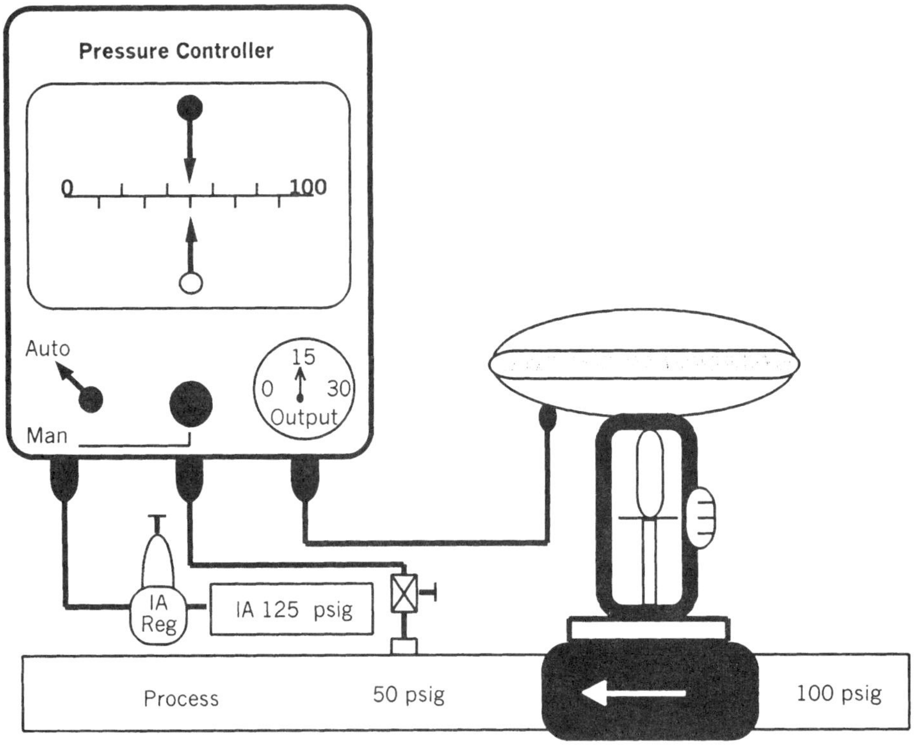

Figure 5.3-3 *Local Controller with Valve*

Figure 5.3-4 *Pressure Controller*

eliminated. Construction and maintenance cost are reduced, but control is not as responsive or precise as conventional transmitter, controller, and positioner control schemes.

5.4 Controller Maintenance

As automatic controllers are designed to compensate for deviations in the control loop, they are seldom responsible for hard failure of the control loop. Provided an accurate input signal and an operable final control element, controllers automatically adjust demand until process is at set-point. The most common pneumatic controller failures are related to inadequate or unclean supply air. Air supply to pneumatic instruments should be routed through a filter regulator. Routine maintenance includes verifying proper regulator adjustment and periodically blowing down the filter trap at the petcock. Pneumatic relays are provided with a restriction cleanout plunger or removable restrictor. Cleanout plungers are affixed to the relay so that pressing the plunger drives a small wire through the fixed restriction and removes any accumulated debris. Removable restrictors are removed and cleaned periodically. Misalignment and wear of the flapper nozzle assembly is also a common failure.

Electronic controllers are extremely dependable devices. The most common failures are associated with the pen and paper drive mechanisms. Most control loop failures are introduced by ancillary devices or cabling problems. Digital controllers rarely fail and control extremely well if programmed properly. Some problems do occasionally occur with the LCD display or keypad.

Chapter Five Summary:

- Controllers provide process variable monitoring (input signal), set-point to process comparison (error signal), and developed correction signal algorithms (output).
- Transmitters and transducers provide the input signal to the controller.
- The controller compares the input signal to a predetermined optimal or desired value, the set-point.
- The controller computes and applies a specific preset algorithm designed to offset or eliminate the error.
- The correction signal is transmitted to a final control element that physically manipulates the controlled variable to effect the correction.
- Controllers monitor inputs continuously, and automatically respond to any changes.
- Proportion band (1/gain × 100) P, integral (Reset) I, and derivative (Rate) D are the three most common response characteristics.
- Proportional values are expressed as percent of scale range, and are generally between 0% and 500%. Proportion band is equal to 1/gain × 100 and gain is equal to 1/proportion band × 100. Proportion band determines response slope relationship between error, in percent of scale, and output, in percent of output range.
- Integral (reset) is a time-weighted response. Integral action develops a correction value determined by the amount of error over time. Integral values are expressed as minutes per repeat or repeats per minute.
- Derivative action is a pre-active correction characteristic that develops correction values relative to the rate of change of the input value.
- Controllers respond to maintain process input value at set-point. The effectiveness of correction is determined by continually comparing process input to set point (feed-back).
- The controller effects all corrective action through a final control element.
- The most common pneumatic signal is 3–15 psig. Pneumatic controllers incorporate a bellows assembly as the input sensor. Bellows induced displacement at one end of the beam is detected as an error signal by the flapper nozzle assembly. The flapper nozzle assembly backpressure is amplified by a pneumatic relay and retransmitted as a pneumatic output signal.
- The relay amplifies the error signal and outputs a representative value of supply air.
- Fulcrum position on the beam determines proportional band through simple lever ratios.
- Reset is developed by a restrictor/tank R/C circuit. The reset tank transmits to a reset bellows located on the beam to compliment the input bellows force.

- A tank circuit also develops rate function. The rate tank, however, accepts input pressure and applies it directly to the output.
- Electronic controllers perform the same basic functions as pneumatic controllers. Although executed by electronic circuits, the control functions of proportional, integral, and derivative are identical.
- A voltage divider circuit on the reference input establishes set-point, and a feed-back resistor on the differential operational amplifier circuit determines gain (proportion).
- An electrical R/C tank composed of a resistor and a capacitor develops integral. A R/C tank also develops derivative.
- Digital controllers offer the ability to configure an array of functions. Digital controllers are also capable of communicating in multi-drop configurations, and performing auto tuning and self-diagnostics. Tuning parameters are accessed through a menu driven keypad or a digital communicator.
- Local controllers perform identical functions as board mount controllers, they are however enclosed in weather tight housings and mounted in the field near the controlled device.
- Local controllers provide local indication of process variable value, set-point adjustment and indication, output indication, manual output adjustment, and adjustment dials for proportional, integral, and derivative.
- Many local controllers provide output balance indicators to assist bumpless transfer functions.
- Automatic controllers are seldom responsible for hard failure of the control loop.
- The most common pneumatic controller failures are related to inadequate, wet, or unclean supply air.
- Routine maintenance includes verifying proper regulator adjustment and periodically blowing down the filter trap at the petcock.
- Pneumatic relays are provided with a restriction cleanout plunger or removable restrictor.
- Misalignment and wear of the flapper nozzle assembly is a common failure.
- The most common electronic controller failures are associated with the pen and paper drive mechanisms.
- Most control loop failures are introduced by ancillary devices or cabling problems.
- Digital controllers rarely fail and control extremely well if programmed properly. Some problems do occasionally occur with the display or keypad.

Chapter Five Review Questions:

1. Describe the basic functions of pneumatic, electronic, and digital controllers.
2. List three operational modes of control.
3. Describe automatic control mode.
4. Describe manual control mode.
5. Describe cascade control mode.
6. Develop the proportion bands (1/gain × 100) P response to a step change in input.
7. Develop the proportion bands (1/gain × 100) P response to a ramp change in input.
8. Develop the integral (Reset) I response to a step change in input.
9. Develop the integral (Reset) I response to a ramp change in input.
10. Develop the derivative (Rate) D response to a step change in input.
11. Develop the derivative (Rate) D response to a ramp change in input.
12. State the purpose of bumpless transfer.
13. State the difference between controller output and actual control valve position in the field.
14. List the functional components of a pneumatic controller.
15. List the control and tuning adjustments available on a pneumatic controller.
16. State the location of control and tuning adjustments on a pneumatic controller.
17. Describe the function of pneumatic controller components including the input bellows, flapper nozzle, force balance beam, feed-back bellows, and output relay.
18. List the functional components of an electronic controller.
19. List the control and tuning adjustments available on an electronic controller.
20. State the location of control and tuning adjustments on an electronic controller.
21. Describe the function of electronic controller components including the voltage divider set-point, error detection amplifier, deviation circuit, integral circuit, and output driver.
22. List the functional components of a digital controller.
23. List the control and tuning adjustments available on a digital controller.
24. State the location of control and tuning adjustments on a digital controller.
25. Describe the function of digital controller components including the analog to digital converter, memory set-point, binary counter, algorithm processor, and digital to analog converter.
26. List the functional components of a local controller.
27. List the control and tuning adjustments available on a local controller.
28. State the location of control and tuning adjustments on a local controller.
29. Describe the function of local controller components.
30. List the common malfunctions associated with pneumatic controllers.
31. List the common malfunctions associated with electronic controllers.
32. List the common malfunctions associated with digital controllers.

Chapter 6

Final Control Elements

OBJECTIVES

Upon completion of Chapter Six the student will be able to:

- *Describe the different control valve types, including spring diaphragm, dual acting piston, motor-operated valves, hydraulic-operated valves, and solenoid valves.*
- *List the basic components of an air-operated control valve.*
- *State the purpose of an I/P transducer.*
- *State the purpose of a positioner.*
- *State the purpose of an actuator.*
- *State the purpose of a valve body.*
- *State the methods to attach a valve assembly to process piping.*
- *State the purpose of the stem packing bonnet.*
- *List several types of packing.*
- *State the purpose of valve stem packing.*
- *State the location of valve packing adjustments.*
- *State the effects of improper packing adjustment.*
- *State three types of valve body trim assemblies.*
- *State the purpose of valve-limit switches.*
- *State the purpose of actuator-mounted solenoids.*
- *State the purpose of boosters.*
- *List the basic components of a motor-operated control valve.*
- *State the purpose of the drive motor.*
- *State the purpose of the torque limiter.*

- *State the purpose of the rotary variable displacement transducer.*
- *State the purpose of the travel limit switches.*
- *List the basic components of a hydraulic-operated control valve.*
- *State the purpose of the servo card.*
- *State the purpose of the servo valve.*
- *State the purpose of the linear variable displacement transformer.*
- *State the purpose of the hydraulic pump.*
- *State the purpose of the hydraulic reservoir.*
- *State the purpose of the accumulator.*
- *List the basic components of a solenoid-operated control valve.*
- *State the purpose of the coil.*
- *State the purpose of the rectifier.*
- *State the purpose of the plunger.*
- *State the preferred method to determine valve operability.*
- *Identify valve performance parameter established by I/P calibration.*
- *Identify valve performance parameter established by positioner calibration.*
- *Identify valve performance parameter established at initial valve calibration.*
- *State technique to determine packing friction.*
- *List malfunctions causing the failure of a valve to open/close.*
- *List malfunctions causing a failure to control (track demand signal).*
- *List malfunctions causing an oscillation.*
- *List malfunctions causing hunting.*
- *List malfunctions causing sticking.*
- *List the affects of control valve malfunction on process parameters.*

6.0 Introduction

Several types of final control elements are available. Control valve types include spring diaphragm, dual acting piston, motor-operated valves, hydraulic-operated valves, and solenoid valves. The operational and control demands of the specific application determine selection. Final control devices are identified by their actuator design and by their body design, Figures 6.0-1 and 6.0-2. The source of motive force employed to position the valve classifies the type of final control element. Air-operated valves (AOVs) convert air pressure into motive force. They are assembled with either spring diaphragm actuators or dual acting piston actuators. AOVs are used in many applications and are quite common. AOV advantages

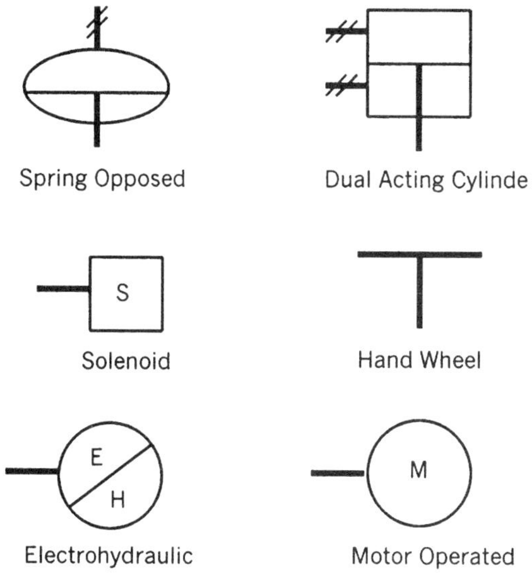

Figure 6.0-1 *Actuator Identification*

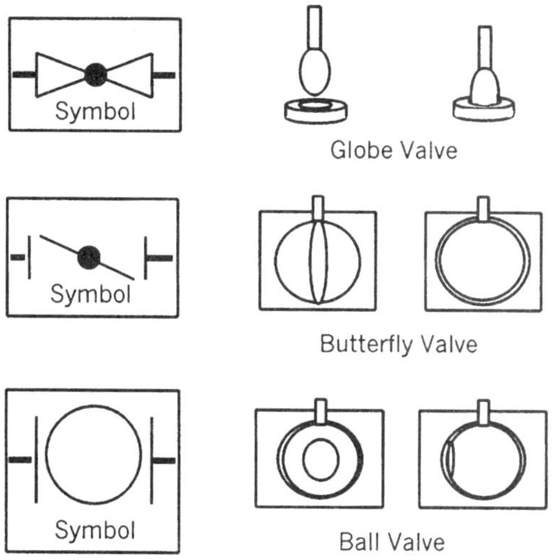

Figure 6.0-2 *Valve Body Identification*

include rapid response and excellent control characteristics. Since typical plant air supply pressures are limited to less than 150 psig, maximum force capabilities are limited by air pressure values. Motor-operated valves (MOVs) and solenoid-operated valves (SOVs) convert electromotive force into valve motion. The MOV actuating device is an electric motor. MOVs generate substantial force, but do not have the rapid response capability of air-operated valves. MOVs are capable of modulated control, but are more commonly used in on/off applications. The SOV actuating device is a coil. SOVs are relatively inexpensive, but energized coils of reasonable size generate minimal force. Hydraulic-operated valves convert hydraulic fluid pressure into valve motion. As typical hydraulic pressures can reach 2000 psig, hydraulic actuators generate extreme amounts of force. Hydraulic valves provide rapid and accurate response. The disadvantages of hydraulic actuators are related to the expense and complexity of the pumps, reservoirs, accumulators, and support devices required.

6.1 Fundamental Components of Final Control Devices

Regardless of the actuator type all valves except solenoid valves have similar body and bonnet assemblies. The bonnet attaches the actuator to the valve body. A packing, or stuffing box, located in the bonnet provides a sealing zone that allows stem motion but prevents process leakage. Several types of packing are available and selection is based on process reactivity, temperature, and pressure. Teflon chevron packing produces the least resistance to valve stem motion, but use is limited to low temperature applications. Braided asbestos packing is used in moderate to high temperatures. Graphite impregnated tape packing is used in extreme temperature applications. Packing does degrade over time and periodic adjustment is required. Packing adjustment nuts are installed on the valve bonnet assembly to facilitate packing adjustment. Self-adjusting constant load packing is available. The

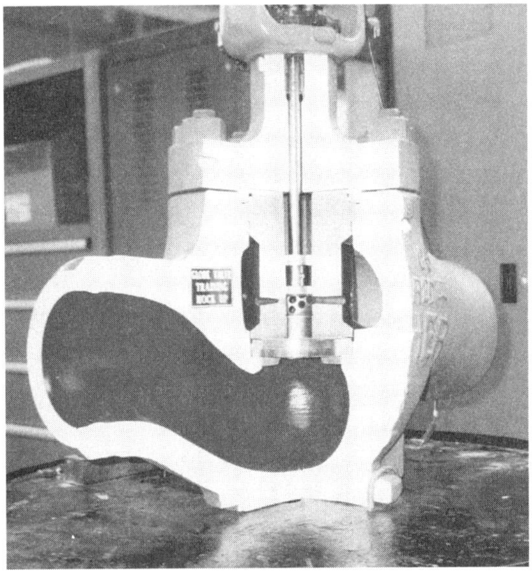

Figure 6.1-1 *Globe Valve Trim (Cutaway)*

Figure 6.1-2 *Travel Limit Position Switches*

Figure 6.1-3 *Control Valve Positioner*

traditional packing nut washers are replaced with a stack of dished (Bellville) washers and torqued to a set value. As the packing collapses under wear the dished washers expand to maintain constant packing torque. Excessive packing friction can affect overall valve performance, including insufficient seat seal off, stem binding, erratic positioning, and reduced stroke time. Loose or worn packing results in process leakage. The valve body encapsulates the valve trim and attaches the valve assembly to the process piping. It provides and maintains the process pressure to environmental sealing boundary at the process connections. Several process piping attachment techniques are employed; including flanged, threaded, and welded.

Valve trim includes all internal process wetted parts, but is specifically the plug, stem, cage, seat, and guides. Three common designs include plug-in seat, piston in cage, and butterfly in window. Plug/seat and piston/cage trims are sliding stem design and butterfly/window are rotary design. Trim sets are physically configured to provide specific flow curves. Linear trim produces a direct linear relationship between valve position and flow. Quick opening trim produces higher flow rates earlier in the valve stroke. Equal percentage trim produces an exponential flow rate in relation to valve position. Butterfly and ball valve trim sets rotate 90 degrees and approximate quick opening flow characteristics. Valve position is indicated by a pointer on the stem that reads out on a scale attached to the valve body. Accessories include limit switches, solenoids, and boosters. Limit switches provide remote indication of valve position and valve open/close logic to program controllers. Solenoids enable and inhibit valve actuation and provide failsafe control of the valve by logic controllers. Boosters multiply air volume from the positioner signal to reduce valve stroke times.

Figure 6.1-4 *Air Supply Regulator and Positioner Feedback Linkage*

6.2 Air-Operated Valves

Air-operated valves (AOVs) are available with spring diaphragm or dual acting piston actuators. The basic components of an air-operated control valve are the actuator, bonnet, body, and trim. Auto/manual controllers generate an analog position demand signal to the air-operated valve. On/off (bi-stable) position demand to air-operated valves is determined by open/close pushbuttons or logic controller actuated solenoids. Limit switches affixed to the valve energize open/close position lights. Many AOVs are supplied with a hand wheel to permit manual operation. Spring diaphragm actuators are assembled with a flexible diaphragm assembly opposing an internal spring. Air pressure acts on the diaphragm to overcome the spring tension and position the valve. When air pressure is removed the internal spring returns the valve to its fail-safe position. Valve travel is limited by diaphragm convolution flexibility, and closing force is limited by diaphragm pressure operating range. Dual acting actuators are assembled as cylinders with an internal piston plate. Air supplied to the top of the piston forces the valve downward, and air supplied to the bottom of the piston forces the valve upward. The valve maintains a fixed position when upper cylinder and lower cylinder air values are equal. With no spring installed, fail-safe position is indeterminate. Typical aluminum cylinders operate at higher-pressure values than diaphragms, and therefore generate additional closing force. The functional components of an air-operated control

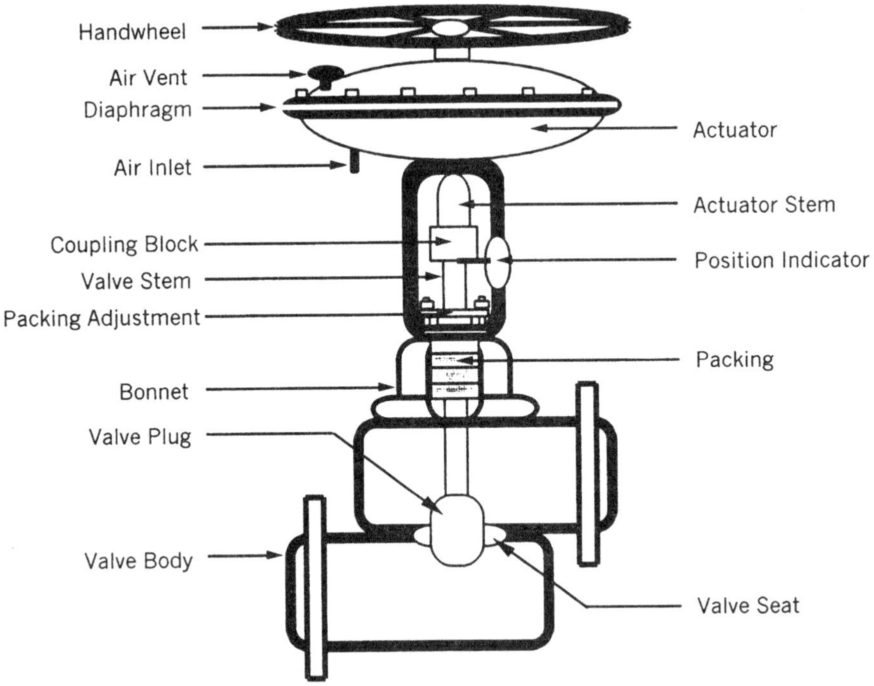

Figure 6.2-1 *Air-Operated Valve*

valve include the actuator, bonnet, body, and trim, and may include an I/P transducer and valve positioner. Accessories include limit switches, solenoids, and boosters. The I/P transducer converts the controller current demand signal into a proportional pneumatic signal

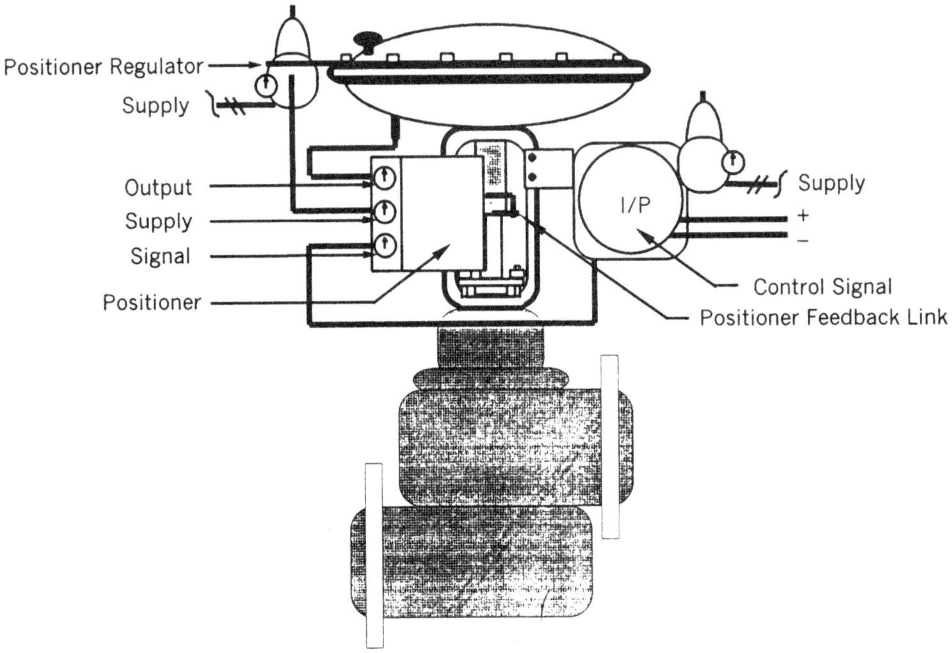

Figure 6.2-2 *Valve with Positioner and I/P*

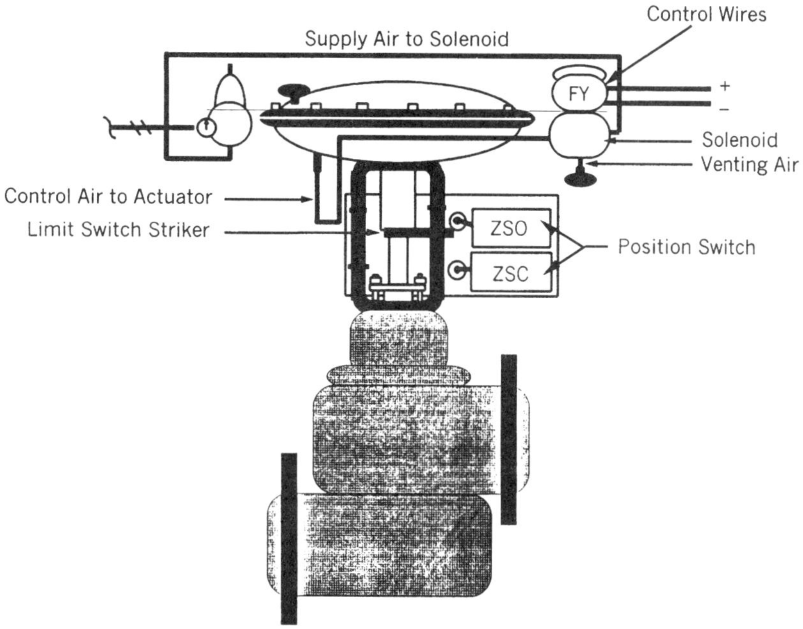

Figure 6.2-3 *Valve with Solenoid and Switches*

and transmits it to the control valve, or positioner if provided. The positioner compares the air signal from the I/P to the valve position and generates an open or close signal to obtain the desired valve position. The positioner-input diaphragm accepts the air signal from the I/P, and the positioner balance beam monitors actual valve position through a feed-back link and cam assembly. The force balance mechanism detects any difference between position demand and actual valve position. The output relay drives a pneumatic signal to the actuator that repositions the valve, restores the beam to a balance condition, and eliminates any difference between demand position and actual position. The actuator receives and amplifies the positioner pressure signal. The actuator develops force as the positioner output pressure is multiplied by the net effective area of the actuator and is converted into linear motion (thrust). The actuator then transmits the thrust/motion to the valve stem through a mechanical coupling. The bonnet assembly attaches the actuator to the valve body. See Figures 6.2-1, 6.2-2, and 6.2-3

Figure 6.2-4 *Air Supply Valve with Positioner and Solenoid*

6.3 Motor-Operated Valves

Motor-operated valve (MOV) position is initiated by operator station pushbuttons on a control panel or by logic controller switches. An internal gear reduction assembly converts rotational motor torque into linear valve motion. Closing force is limited by current capacity of the motor control. Internal travel limit switches that energize open/close position indication lamps determine valve position indication, and travel-limit switches that enable or interrupt motor voltage implement valve travel control. Motor controlled valves may be supplied with hand wheels to permit valve operation in the event of power failure or motor malfunction. An engagement clutch is used to select manual or motor operation. The functional components of a motor-operated control valve are the drive motor, gear reducer, torque limiter, rotary variable displacement transducer (RVDT), travel-limit switches, position indication switches, and valve body. The drive motor is a high-torque electric motor. The gear reducer assembly converts motor rotation into linear valve stem travel. The torque limiter is a clutched disengagement assembly designed and adjusted to protect the valve trims from damage. The RVDT is a precision rotation detector; it is used to provide feed-back for analog operation of rotary travel valves. Travel limit-switches interrupt voltage to the motor at preset valve travel limits to prevent over torque and excessive current demand. Position indication switches energize and illuminate open and closed position lights and provide valve position logic to program controllers.

Figure 6.3-1 *Motor-Operated Valve*

The functional components of a solenoid-operated control valve include the coil, rectifier, plunger, and valve body. The coil provides motive force to position the plunger. The rectifier converts AC voltage into DC voltage. The plunger changes position to redirect process or air flow through the solenoid. Solenoid valves are available in DC and AC configurations. Solenoid activation is initiated through operator-controlled pushbuttons, pressure, flow, or temperature switches, or program logic controllers. Solenoid valves are most commonly used to affect on/off control of air-operated valves or in lube oil trip circuits on compressors and turbines. The common tubing is attached to the actuator and solenoid actuation selects between supply air and atmospheric vent, or between normal flow and trip flow.

6.4 Hydraulic-Operated Valves

Hydraulic-operated valves are common in applications requiring precise control of large flow rates. Hydraulic valve position demand is generated through a servo-control unit or panel mounted operator push buttons. Linear variable displacement transformers (LVDTs) provide position feed-back, and limit switches illuminate open/close lights. Typical installations are turbine driven pumps and generators. Functional components of a hydraulic-operated control valve include the actuator, servo valve, servo card, LVDT, hydraulic supply system, and valve body. Control valve action is initiated through pushbuttons or speed control devices. Limit switches are often installed to provide position indication and logic. The actuator is a heavy-duty steel cylinder with an internal piston. Hydraulic fluid pressure applied above the piston provides downward force, and fluid pressure applied below the piston provides upward force. Most hydraulic actuators have extremely large springs that provide stability and fail safe assist. The servo valve is attached between the hydraulic fluid supply and drain lines and the actuator. The servo valve alternately directs hydraulic fluid supply pressure to one side of the piston while draining fluid from the other side. The servo valve can also hold fixed pressure across the piston to stop the valve at mid position. Figures 6.4-1, 6.4-2, and 6.4-3 show three conditions of operation. Servo valve bias is adjusted to fail the hydraulic valve to the fail-safe position. The servo card is usually mounted in a remote controller unit. It compares position demand to actual valve position and transmits a correction signal to the servo valve. The LVDT linear variable differential transformer is attached to the hydraulic cylinder and valve stem. It converts valve position into a voltage signal and transmits that information to the servo card.

The hydraulic supply system in Figure 6.4-4 consists of a high-pressure pump, a reservoir, an accumulator, and various heaters, coolers, and filters. The hydraulic pump supplies high-pressure fluid to the valve. The hydraulic reservoir provides a safe, clean, controlled supply of fluid to the pump. The accumulator provides an inert gas-blanketed, remote reservoir of hydraulic fluid to the system. The accumulator assists in maintaining regulated pressures throughout the system. It increases overall system response and reduces cycling of the pump.

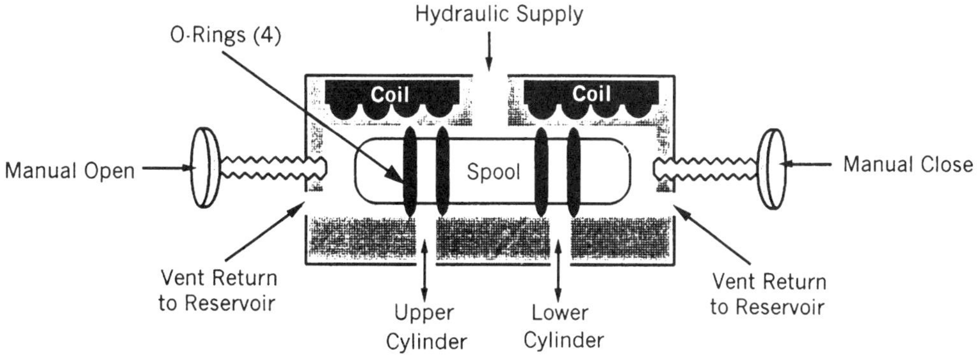

Figure 6.4-1 *Hydraulic Shuttle Valve (Valve Stable)*

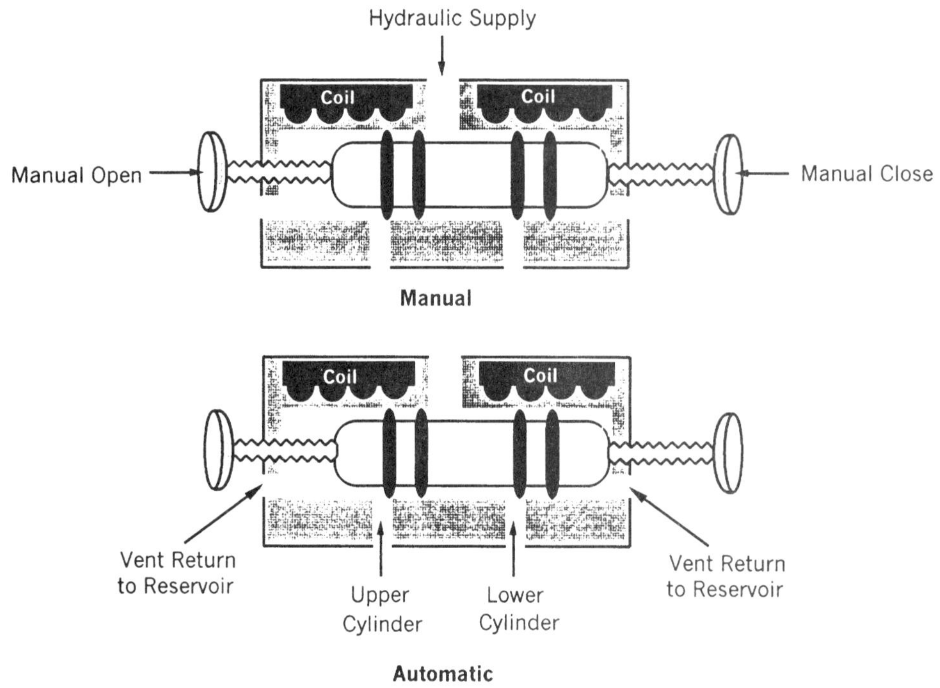

Figure 6.4-2 *Hydraulic Shuttle Valve (Valve Opening)*

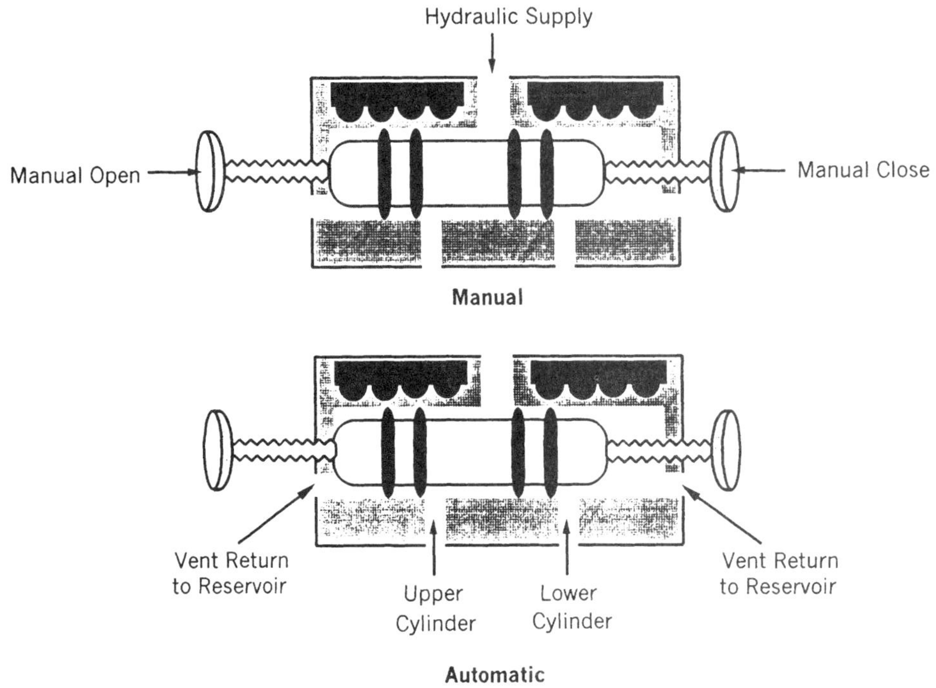

Figure 6.4-3 *Hydraulic Shuttle Valve (Valve Closing)*

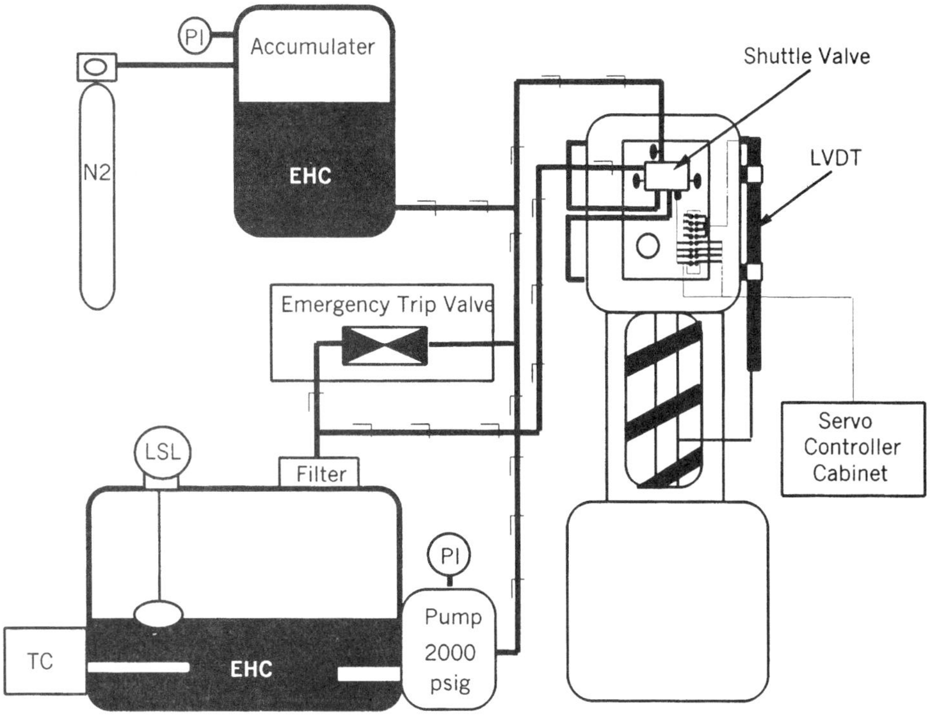

Figure 6.4-4 *Electro-Hydraulic Valve Support System*

Figure 6.4-5 *Electro-Hydraulic Control Oil Skid*

Proper maintenance of electro-hydraulic control systems includes periodic filter replacement, reservoir level verification, reservoir vent desiccant replacement, accumulator blanket charging, visual leak inspection, and hydraulic fluid chemistry monitoring.

6.5 Troubleshooting Final Control Devices

Several methods are available to identify control valve malfunction. The most conclusive technique to determine valve operability is to stroke the valve full range with the controller, and compare controller output signal to positioner input signal, and/or valve position at several points. Verify that positioner input is proportional to controller output. These values may be established by the I/P transducer calibration. The positioner output, should begin to ramp when the input value is equivalent to the lower calibration range (typically 3.25 psig). The positioner output should saturate when the input value is equivalent to the upper calibration range (typically 14.75 psig). To compare positioner output to valve position, verify the valve begins to stroke at the positioner output value equivalent to the valve's lower spring set value, then verify valve travel is full stroke at the positioner output value equivalent to the valve's upper spring set value. These values are established at initial valve calibration and rarely, if ever, change. Finally, compare the actuator air pressure value required to position the valve at mid-stroke from both increasing and decreasing directions. The difference in pressure values multiplied by the net effective actuator diaphragm area is equivalent to the frictional resistance of the valve. Excessive friction indicates binding or internal damage.

Failure to open or close is usually due to a loss of demand signal (I/P), loss of supply air (regulator), actuator failure, or trim failure. Failure to control (track demand signal) is usually caused by water in the airline, insufficient air supply pressure or volume, I/P out of calibration, positioner out of calibration, or valve specifications inappropriate for the application. Excessive positioner or booster gain, a worn, loose, or binding feed-back link, or a loose positioner-mounting bracket induces oscillation. Hunting is often caused by the positioner input bellows leaking signal air, water in the instrument airline, insufficient supply pressure or volume, positioner gain to low, or the actuator diaphragm leaking by. Sticking is the result of excessive packing friction or internally damaged components.

Control valves are primarily used as final control elements. As such, any loss of control related to the valve results in a representative loss of control over the process. Control valve malfunctions can result in poor quality, low yield, damaged equipment, environmental contamination, personal injury, or loss of human life.

Chapter Six Summary:

- Control valve types include spring diaphragm, dual acting piston, motor-operated valves, hydraulic-operated valves, and solenoid valves.
- The basic components of an air-operated control valve are the actuator, bonnet, body, and trim.
- The I/P transducer converts controller current demand into a proportional pneumatic signal and retransmits it to the positioner.
- The positioner accepts position demand signal from the I/P and drives a pneumatic signal to the actuator that repositions the valve.
- The actuator receives positioner output pressure converts it into linear motion (thrust), then transmits the thrust/motion to the valve stem through a mechanical coupling.
- The valve body encapsulates the valve trim, attaches the valve assembly to the process piping, and provides a sealing boundary at the process mounting points.
- The bonnet attaches the actuator to the valve body
- Common types of packing are Teflon chevron, braided asbestos rope, graphite-impregnated tape, and self-adjusting graphite.
- Valve stem packing in the bonnet provides a sealed orifice that allows stem motion but prevents process leakage.
- Packing degrades over time and periodic adjustment is required. Packing adjustment nuts are installed on the valve bonnet assembly to facilitate packing adjustment.
- Excessive packing friction results in insufficient seat seal off, stem binding, erratic positioning, and reduced stroke time. Loose or worn packing results in process leakage.
- Three common body designs include plug-in seat, piston in cage, and butterfly in window.
- Limit switches provide remote indication of valve position and valve open/close logic to program controllers.
- Solenoids enable and inhibit valve actuation and provide fail-safe control of the valve by logic controllers.
- Boosters multiply air volume from the positioner signal to reduce valve stroke times.
- The functional components of a motor-operated control valve are the drive motor, gear reducer, torque limiter, rotary variable displacement transducer (RVDT), travel limit switches, position indication switches, and valve body.
- The drive motor is a high torque electric motor that provides force to the gear reducer assembly where it is converted into valve stem travel.
- The torque limiter is a clutched disengagement assembly designed and adjusted to protect the valve trim, gear reducer, and motor from damage.
- Travel-limit switches interrupt voltage to the motor at preset valve travel- limits to prevent over torque and excessive current demand.

- Functional components of a hydraulic-operated control valve include the actuator, servo valve, servo card, linear variable displacement transformer (LVDT), hydraulic supply system, and valve body.
- The servo card compares position demand to actual valve position and transmits a correction signal to the servo valve.
- The servo valve alternately directs hydraulic fluid supply pressure to one side of the piston while draining fluid from the other side to position the valve, or holds a fixed pressure across the piston to stop the valve at mid position.
- The accumulator assists in maintaining regulated pressures throughout the system. It increases overall system response and reduces cycling of the pump.
- Proper maintenance of electro-hydraulic control systems includes periodic filter replacement, reservoir level verification, reservoir vent desiccant replacement, accumulator blanket charging, visual leak inspection, and hydraulic fluid chemistry monitoring.
- The functional components of a solenoid-operated control valve include the coil, rectifier, plunger, and valve body.
- The coil provides motive force to position the plunger.
- The rectifier converts AC voltage into DC voltage.
- The plunger changes position to redirect process or air flow through the solenoid.
- To determine valve operability stroke the valve full range with the controller, and compare controller output signal to positioner input signal, and valve position at several points.
- I/P calibration establishes the positioner input proportional to controller output.
- Positioner calibration establishes the point that positioner output begins to ramp at the positioner input value equivalent to the lower calibration range (typically 3.25 psig), and the point that positioner output saturates as positioner input value equivalent to upper calibration range (typically 14.75 psig).
- Failure to open or close is usually due to a loss of demand signal (I/P), loss of supply air (regulator), actuator failure, or trim failure.
- Failure to control (track demand signal) is usually caused by water in the airline, insufficient air supply pressure or volume, I/P out of calibration, positioner out of calibration, or valve specifications inappropriate for the application.
- Sticking is the result of excessive packing friction or internally damaged components.
- Unfavorable results of control valve malfunction include poor quality, low yield, equipment damage, environmental contamination, personal injury, or loss of human life.

Chapter Six Review Questions:

1. Describe the different control valve types, including spring diaphragm, dual-acting piston, motor-operated valves, hydraulic-operated valves, and solenoid valves.
2. List the basic components of an air-operated control valve.
3. State the purpose of an I/P transducer.
4. State the purpose of a positioner.
5. State the purpose of an actuator.
6. State the purpose of a valve body.
7. State three methods to attach a valve assembly to process piping.
8. State the purpose of the stem-packing bonnet.
9. List four types of packing.
10. State the purpose of valve stems packing.
11. State the location of valve packing adjustments.
12. State the effects of improper packing adjustment.
13. State three types of valve body trim assemblies.
14. State the purpose of valve-limit switches.
15. State the purpose of actuator-mounted solenoids.
16. State the purpose of boosters.
17. List the basic components of a motor-operated control valve.
18. State the purpose of the drive motor.
19. State the purpose of the torque limiter.
20. State the purpose of the RVDT.
21. State the purpose of the travel-limit switches.
22. List the basic components of a hydraulic-operated control valve.
23. State the purpose of the servo card.
24. State the purpose of the servo valve.
25. State the purpose of the LVDT.
26. State the purpose of the hydraulic pump.
27. State the purpose of the hydraulic reservoir.

28 State the purpose of the accumulator.

29. List the basic components of a solenoid-operated control valve.

30. State the purpose of the coil.
31. State the purpose of the rectifier.
32. State the purpose of the plunger.
33. State the preferred method to determine valve operability.
34. Identify valve performance parameter established by I/P calibration.
35. Identify valve performance parameter established by positioner calibration.
36. Identify valve performance parameter established at initial valve calibration.
37. State technique to determine packing friction.
38. List malfunctions causing a failure to open/close malfunction.
39. List malfunctions causing a failure to control (track demand signal).
40. List malfunctions causing an oscillation.
41. List malfunctions causing hunting.
42. List malfunctions causing sticking.
43. List the affects of control valve malfunction on process parameters.

Chapter 7

Emergency Shutdown & Interlock Systems

OBJECTIVES

Upon completion of Chapter Seven the student will be able to:

- *State the purpose and intent of emergency shutdown systems (ESDs).*
- *List the fail-safe modes of a valve.*
- *Identify a fail-open valve.*
- *Identify a fail-close valve.*
- *Identify a fail-as-is (in position) valve.*
- *State the purpose of fail-safe valves.*
- *List five fail-safe support devices.*
- *Define the term single point failure.*
- *Define the term mitigate.*
- *State the fundamental requirement of a fail-safe system.*
- *Describe several methods used by ESD systems for securing the process.*
- *Define redundancy.*
- *State the benefits of redundant controls.*
- *State the drawbacks of redundant controls.*
- *State three methods used to determine if redundant control is required.*
- *State the purpose of interlocks, and permissives.*
- *State the purpose of trips.*
- *List six process variables typically monitored by interlocks, trips, or permissives.*

- *List the machine variables typically monitored as interlocks, trips, or permissives.*
- *List the motive medias used to effect interlocks, trips, and permissives.*

7.0 Introduction

The purpose and intent of ESD (emergency shutdown) systems is to; eliminate unnecessary trips or transients due to single point failures, reduce or mitigate undesirable effects of human performance errors, provide a safe and controlled response to unexpected conditions, protect equipment from damage, protect the environment from contamination, and prevent personal injury. There are many engineered safety features designed into critical systems to secure the process. The extent of safeguards incorporated into a particular system is determined by the inherent volatility of the process and the negative impact potential resulting from a loss of control. One or more techniques are employed, based on engineering review of design-based impact assessments, industry events, and plant operating experience.

7.1 Redundancy

Installation of redundant instrumentation is an expensive but extremely effective safeguard solution. Several identical channels simultaneously monitor the same process parameter. In Figure 7.1-1, each channel monitors the process as well as the other companion channels. If any single channel generates a significantly different reading, it is flagged as a deviant channel. The suspect channel is then automatically removed from service. The process

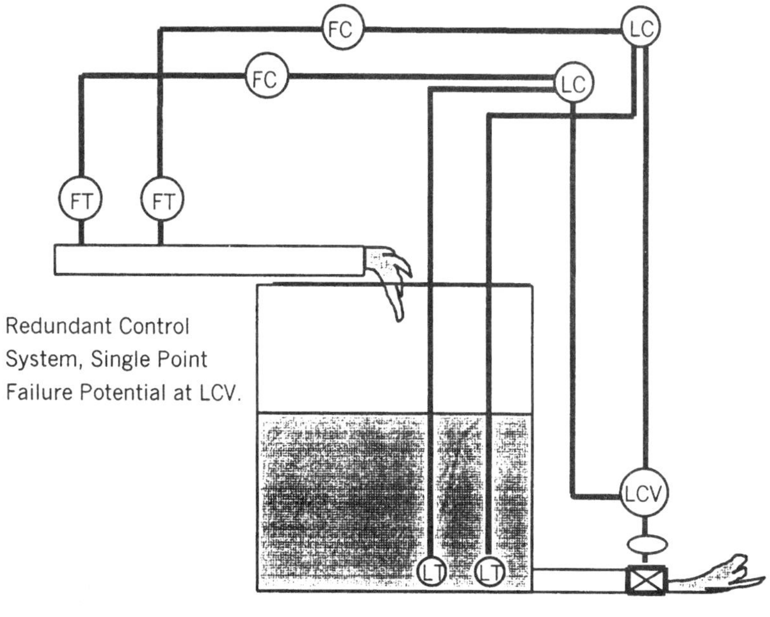

Figure 7.1-1 *Redundant Safeguard*

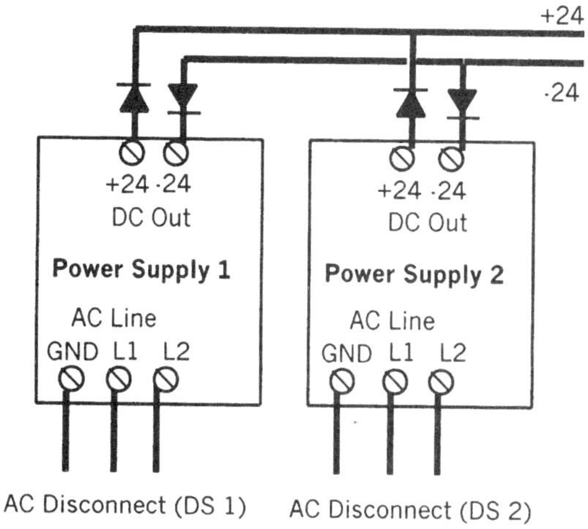

Figure 7.1-2 *Redundant Power Supplies*

remains unaffected and continues to operate normally on the remaining channels. As an added benefit, single channels are removed from service to support on-line maintenance and testing without affecting overall system safety or response. A dual power supply system only loads a single power supply unit, Figure 7.1-2. If one power supply fails in service, the other instantly accepts load. The unloaded power supply is replaced on-line without interruption. Sourcing AC line power from different disconnects abates single point failure due to switchgear failure or operator error. Figure 7.1-3 shows dual redundant trip solenoids powered from separate sources and connected to the valve with stainless steel flexible hose.

Figure 7.1-3 *Redundant Trip Solenoids*

7.2 Fail-safe Design

Properly designed fail-safe systems are not dependent on external energy sources to default the configuration of final control elements to a safe position. A loss or interruption of electrical, hydraulic, or pneumatic control power initially transfers critical components to a redundant power source. If the back-up power source fails to actuate, critical components default to their shelf (de-energized) state and initiate a safe shutdown sequence. Control valves are configured for specific fail-safe positions and particular failure modes. Failure mode selections include fail-as-is, fail-open, or fail-closed. Visual inspection can often determine the fail mode. Fail-open valves require air to close on top of the diaphragm, fail-close valves require air to open on the bottom of the diaphragm, and fail-as-is valves accept air to the top and bottom of the diaphragm. Fail modes are designed to configure or restore the process to the safest or most desirable condition in the event of a loss or degradation of control system capabilities. Additional failsafe support devices include solenoids, springs, accumulators, and shuttle valves.

7.3 Interlocks, Trips, and Permissives

The purpose of interlocks, trips, and permissives is to automatically interrupt or reconfigure final control device(s) if monitored variables significantly deviate from specifications. They allow equipment to start and operate only when monitored variables are within designed parameters, Figure 7.3-1. Typical process variables monitored are; process flow, level, temperature, pressure, chemistry, contamination, and reactivity. Typical machine variables monitored are coolant level, temperature, and pressure; lubricant level, temperature, and pressure; and machine vibration, temperature, and speed. Trips automatically de-energize or turn off equipment if monitored variables exceed preset values, Figure 7.3-2. Interlocks and permissives inhibit unanticipated actuation of equipment and ensure correct start-up/shutdown sequences are followed. Interlock designs are mechanical, hydraulic, pneumatic, or electrical. Electrical interlocks are either (relay logic), hard-wired or (processor controlled), software.

Figure 7.3-3 depicts a mechanical over-speed trip device. This device is common on large rotating equipment. The trip plunger is recessed into a cylindrical bore perpendicular to the shaft. A counter spring is installed to a specific compression to resist the plunger inertia at normal operating speeds. If machine speed exceeds the desired limit, the plunger generates sufficient inertia to compress the spring. As the spring compresses, the plunger protrudes from the bore and actuates the trip lever. As in Figure 7.3-4, the mechanical over-speed and manual trip pushbutton often trigger the same trip lever. This particular arrangement also functions as a permissive. The turbine receives energy to turn from the boiler steam

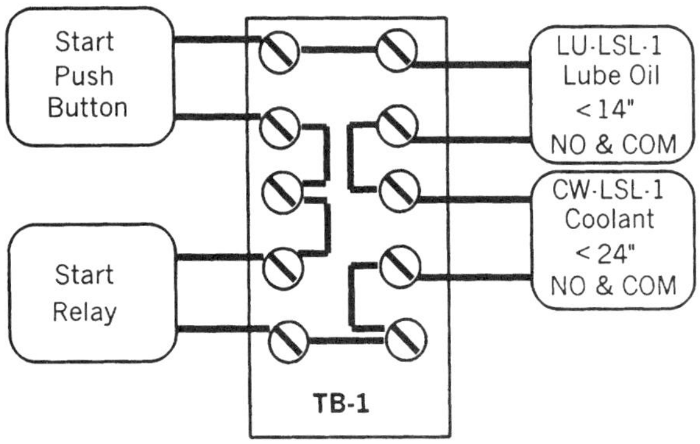

Figure 7.3-1 *Hard-Wired Permissive String*

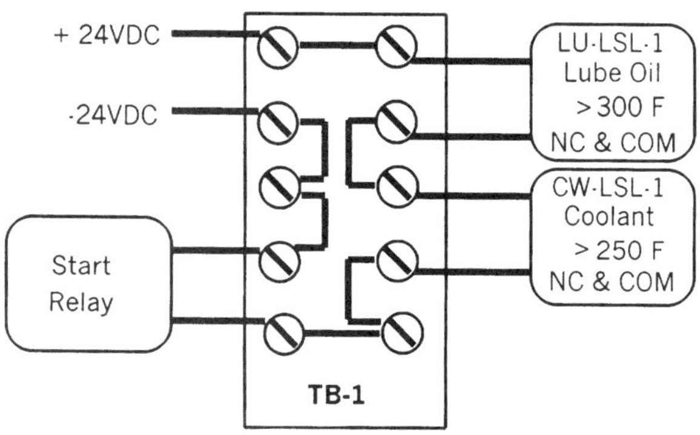

Figure 7.3-2 *Hard-Wired Trip String*

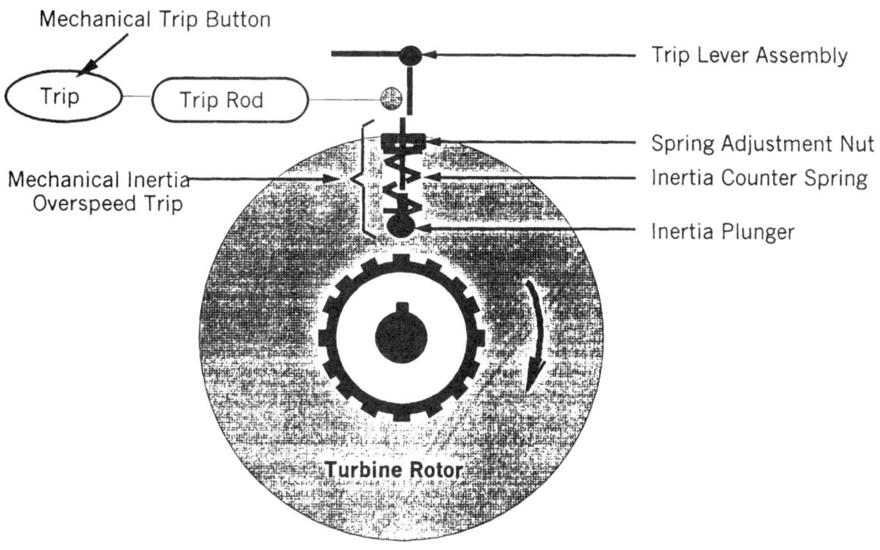

Figure 7.3-3 *Mechanical Trip Assembly*

header. Hydraulic-operated throttle valves control the amount of steam admitted to the turbine. The throttle valves are designed to fail-closed and stop all steam from entering the turbine on a loss of control oil. When in the open position, the interlock valve drains control oil back to the reservoir, and prevents the control oil header from pressurizing. The interlock valve must close to build enough control oil header pressure to open the throttle valves. The interlock valve is similar to a fail-open, air-operated valve, except instead of using air

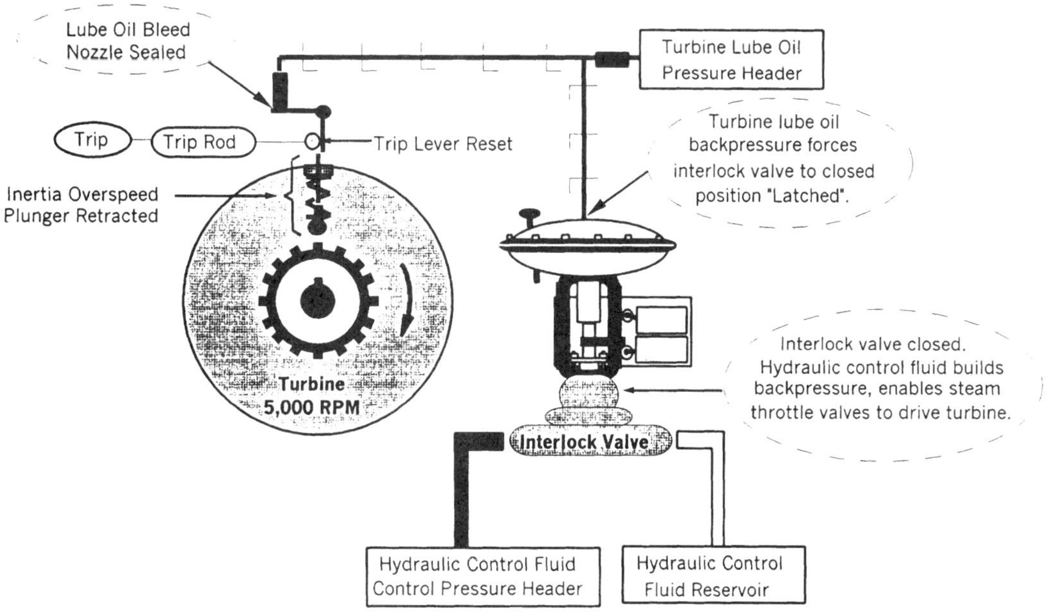

Figure 7.3-4 *Overspeed Reset/Turbine Latched*

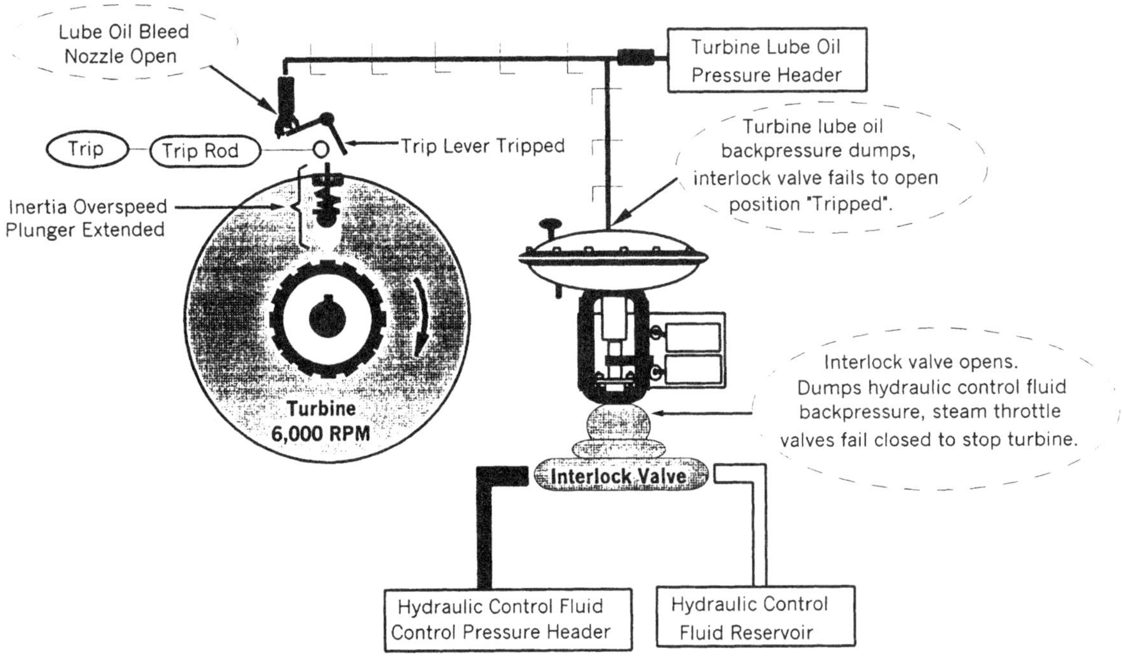

Figure 7.3-5 *Overspeed Tripped / Turbine Tripped*

to close, lubricating oil pressure is applied to the actuator. Therefore the interlock valve prohibits turbine operation unless adequate lubricating oil pressure is available. In Figure 7.3-5, a restrictive orifice supplies bearing lubricating oil to the intercept valve and the trip lever bleed nozzle. The orifice is sized to provide less oil than the bleed can vent. If the turbine over-speeds the plunger extends, engages the trip lever, and vents bearing lubricating oil from the interlock valve actuator. As pressure at the actuator decreases the interlock valve opens and depressurizes the control oil header. Once control oil pressure drops, the throttle valves fail close, the trip/throttle valves shut-off steam to the turbine, and the turbine stops.

Figure 7.3-6, displays additional redundancy attained by installing a true dual-redundant electrical over-speed trip system. The system monitors turbine speed by electronically sensing the frequency of pulses produced by a toothed wheel attached to the turbine shaft. The speed controller compares actual speed to maximum allowable speed-set point. If either speed probe senses excess speed, the controller interrupts electrical power to both solenoid dump valves. One solenoid fails open and dumps bearing oil from the interlock valve actuator. This forces the interlock valve to fail-open and dump control oil pressure. The other solenoid fails-open and dumps control oil directly to the reservoir. This type of system is

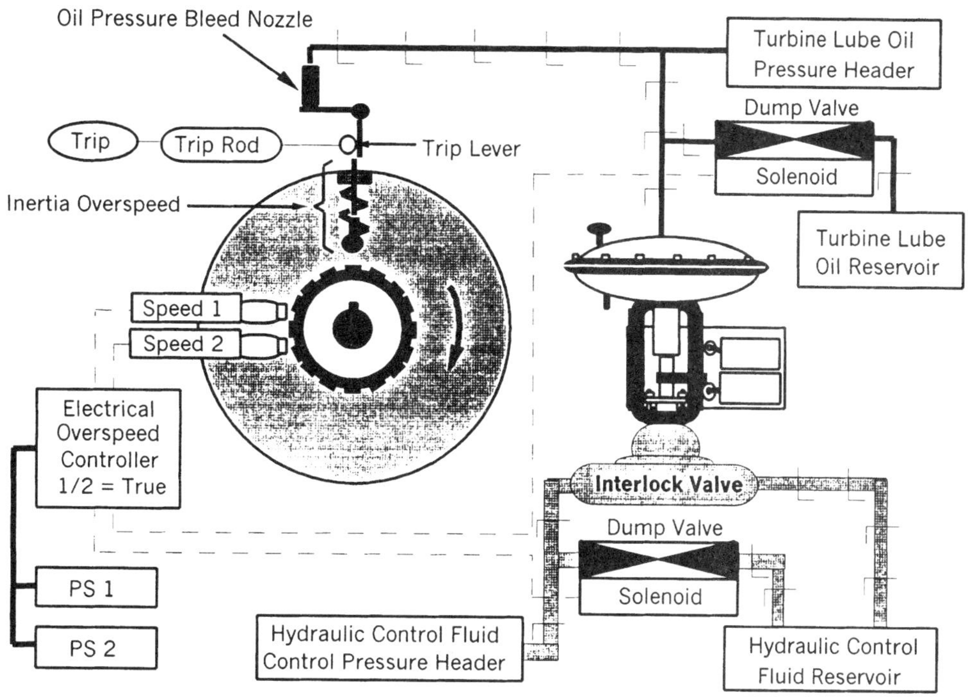

Figure 7.3-6 *Triple Redundant Overspeed Protection*

extremely reliable, especially if true redundant power supplies are provided to the electrical over-speed controller. A magneto is sometimes attached to the machine's shaft as a permanent generator to provide electrical power to the control circuits as long as the shaft is spinning. Some turbines have shaft-driven lubricating oil pumps that provide the primary bearing lubricating oil. In this case, the back-up lubricating oil pump is electrically driven and only used during start-up or if the primary shaft driven pump fails. The safeguards applied in this example are fairly extensive, they are expensive to design, install, and maintain. This is however, a realistic representation considering the destructive forces involved in the obliteration of a large turbine due to over-speed.

Chapter Seven Summary:

The purpose and intent of emergency shutdown systems is to:

- Eliminate unnecessary trips or transients due to single point failures,
- Reduce or mitigate undesirable effects of human performance errors,
- Provide a safe and controlled response to unexpected conditions,
- Protect equipment from damage,
- Protect the environment from contamination, and
- Prevent personal injury.
- Installation of redundant instrumentation is used by ESD systems for securing the process.
- Control valves are configured for specific failsafe positions and particular failure modes.
- Failure mode selections include fail-as-is, fail-open, or fail-closed.
- Fail-open valves require air to close on top of the diaphragm, fail-close valves require air to open on the bottom of the diaphragm, and fail-as-is valves accept air to the top and bottom of the diaphragm.
- Fail modes are designed to configure or restore the process to the safest or most desirable condition in the event of a loss or degradation of control system capabilities.
- Additional failsafe support devices include solenoids, springs, accumulators, and shuttle valves.
- Redundant systems use several identical hardware channels to simultaneously monitor the same point.
- All channels monitor the process and the other companion channels.
- The system-cross comparisons flag the deviant channel.
- The suspect channel is automatically removed from service.
- The system remains unaffected and continues to operate normally on the remaining channels.
- As an added benefit single channels are removed from service to support on-line maintenance and testing without affecting system safety or response.
- The extent of safeguards incorporated is determined by the volatility of the process and the negative impact resulting from a loss of control.
- One or more techniques are employed, based on engineering review of design based impact assessments, industry events, and plant operating experience.
- True failsafe systems return the process to a controlled condition without dependency on electrical, pneumatic, or hydraulic power.
- Trips automatically interrupt or reconfigure final control devices if the measured variable exceeds specifications.

- The purpose of interlocks and permissives is to allow equipment to start and operate only when monitored variables are within designed parameters.
- Typical process variables monitored by interlocks, trips, and permissives are flow, level, temperature, pressure, chemistry, contamination, and reactivity.
- Typical machine variables monitored by interlocks, trips, and permissives are coolant level, temperature, and pressure; lubricant level, temperature, and pressure; and machine vibration, temperature, and speed.
- Interlock designs are mechanical, electrical hard-wired (relay logic), electrical software (processor controlled), pneumatic, or hydraulic.

Chapter Seven Review Questions:

1. State the purpose and intent of emergency shutdown systems.
2. List three fail-safe modes of a valve.
3. Identify a fail-open valve.
4. Identify a fail-close valve.
5. Identify a fail-as-is valve.
6. State the purpose of fail-safe valves.
7. List the fail-safe support devices.
8. Define the term single point failure.
9. Define the term mitigate.
10. State the fundamental requirement of a fail-safe system.
11. Describe several methods used by ESD systems for securing the process.
12. Define redundancy.
13. State the benefits of redundant controls.
14. State the drawbacks of redundant controls.
15. State three methods used to determine if redundant control is required.
16. State the purpose of interlocks and permissives.
17. State the purpose of trips.
18. List the process variables typically monitored by interlocks, trips, or permissives.
19. List the machine variables typically monitored as interlocks, trips, or permissives.
20. List the motive medias used to effect interlocks, trips, and permissives.

Chapter 8

Programmable Logic Controllers

OBJECTIVES

Upon completion of Chapter Eight the student will be able to:

- *State the purpose and function of a programmable logic controller (PLC).*
- *List the fundamental hardware components of a PLC.*
- *State the function of the individual hardware components of a PLC.*
- *List the typical inputs/outputs monitored by a PLC.*
- *Convert binary numbers into octal and hexadecimal equivalents.*
- *Identify ladder logic symbols.*
- *List typical logic operands used by a PLC.*
- *Interpret ladder logic diagrams.*
- *Identify the hardware termination input/output point specified by the ladder logic.*
- *Determine if the anticipated field input is recognized by the PLC.*
- *Determine if a module contains a faulty fuse.*
- *Determine if the interface module is communicating.*
- *Determine if the power supply is O.K.*
- *Locate and interpret self-health status monitoring LEDs.*
- *State the conditions required to generate a discordance alarm.*

8.0 Introduction

Programmable logic controllers (PLCs) combine extremely high computation ability, diverse flexibility, and excellent reliability into a space efficient control system. Their use is widespread throughout industry. PLCs control livestock feed and watering systems, lathes and milling machinery, water purification, and a host of assembly line operations. Batch processes in chemical, pharmaceutical, and food/beverage industries all rely on PLCs. Many skid mounted process systems are delivered with preprogrammed PLCs already installed. Since PLCs have the ability to communicate with other processors, they are easily added to existing systems. Since they can be reprogrammed, they adapt readily to design alterations and can be reconfigured as required to meet emerging needs.

8.1 Programmable Logic Controller Hardware

Programmable logic controllers are microprocessor based sequential step software configurable controllers. Typical hardware, shown in Figure 8.1-1, consists of a mounting rack, a power supply, a communication interface module, a processor module (CPU), and a selection of user-defined plug in input/output modules. The mounting rack provides a stable mounting platform for the modules. It also contains the back-plane bussing network that distributes power and data through the rack. The power supply provides back-plane power to the modules and signal power, or wetting voltage, for the output modules. The communication

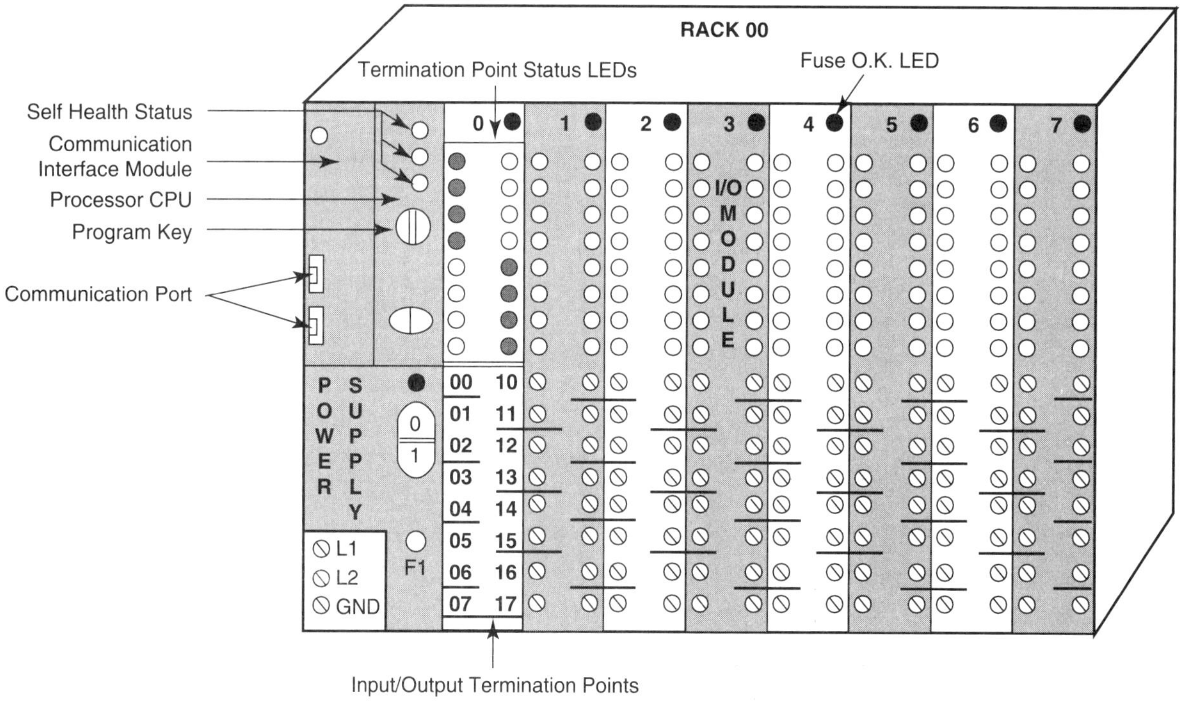

Figure 8.1-1 *Programmable Logic Controller Hardware*

interface module provides data transport links to other racks in the system, and RS232 serial or RJ11 phone connections to facilitate user programming with a laptop computer or hand held communicator. The processor module (CPU) provides memory locations for executive programs, application programs (ladder logic), data bits, and registers. The processor performs all logic functions, and executes the application program. Input/output modules receive and transmit data to and from field devices. They provide electrical isolation to protect the CPU and transpose data into the appropriate retransmission format. The type of data transmitted determines input/output module selection. Specific modules are selected to input or output either binary or analog data. Typical input modules are installed to monitor; AC volts binary input, DC volts binary input, DC volts analog input, DC current analog input, thermocouple analog input, and RTD analog input. Typical output modules are AC volt binary, DC volt binary, and DC volts analog. Multiple rack configurations are capable of sharing input data through the communication interface module. The PLC monitors external contacts, performs logical computation and comparison to the user installed program code, and outputs the desired data. The installed program code is accessible through the communication module, and is revised or updated as required. The program software is presented in a user friendly, ladder diagram format.

8.2 Binary Numbering Systems

Programmable logic controllers only manipulate data in binary format. Since long strings of ones and zeros are difficult for people to work with, alternate numbering systems are used to represent groups of three and four bits of binary data as alphanumeric symbols. The alternate numbering systems used are binary (base two), octal (base eight), and hexadecimal (base sixteen). Figure 8.2-1 lists the equivalent values of these numbering systems through decimal 15. To avoid confusion multiple based numbering systems are identified by subscript: $decimal_{10}$, $binary_2$, $octal_8$, and $hexadecimal_{16}$. Since multi-digit non-decimal numbers represent different values than the decimal value of the same symbols, different terms are used. The octal value 13_8 is referred to as one three octal, not thirteen. The hexadecimal term 13_{16} is referred to as one three hex. Additionally, when raising numbers to powers or using exponents, any number raised to the power of zero is one. Any number raised to the power of one is unchanged. Numbers raised to powers other than zero (0) or one (1) are multiplied by themselves the number of times indicated by the power to which they are raised. Some examples are $2^0 = 1$, $2^1 = 2$, $2^2 = 4$, and $10^3 = 1000$.

The universal numbering system in use today is the decimal system. The decimal system is base ten. This means there are only ten symbols to select from. The symbols are 0, 1, 2, 3, 4, 5, 6, 7, 8, and 9. These ten symbols are arranged in sequence to represent any numeric value. This works well as the count progresses from 0 to 9. But, when the count reaches 9, all available symbols are used. To progress the count, place weighting is required. A 0 is used to indicate the one's column is full and the count is resumed in the ten's column. Therefore any decimal number is actually a summation of the counts in each column, with each column representing an additional multiple of ten.

Decimal	Binary (Four Bit)	Octal	Hexadecimal
0	0000	0	0
1	0001	1	1
2	0010	2	2
3	0011	3	3
4	0100	4	4
5	0101	5	5
6	0110	6	6
7	0111	7	7
8	1000	10	8
9	1001	11	9
10	1010	12	A
11	1011	13	B
12	1100	14	C
13	1101	15	D
14	1110	16	E
15	1111	17	F

Figure 8.2-1 *Numerical Conversion*

Example: the number 1942 represents,
A count of "1" in the one thousand column $(1 \times 10^3) = 1 \times 1000 = 1000$
Plus a count of "9" in the one hundred column + $(9 \times 10^2) = 9 \times 100 = 900$
Plus a count of "4" in the ten column + $(4 \times 10^1) = 4 \times 10 = 40$
Plus a count of "2" in the one column + $(2 \times 10^0) = 2 \times 1 = 2$
The summation of these counts is 1942_{10}.

Decimal, binary, octal, and hexadecimal numbering systems all adhere to the same place weighting principle.

The binary system is base two. This means there are only two symbols to select from. The symbols are 0 and 1. These two symbols are arranged in sequence to represent any numeric value. This works well as the count progresses from 0 to 1. But, when the count reaches 1, all available symbols are used. To progress the count, place weighting is required. A "0" is used to indicate the one's column is full and the count is resumed in the two's column. Therefore, any binary number is actually a summation of the counts in each column, with each column representing an additional multiple of two.

Example: the number 1101_2 represents,
A count of "1" in the eight column = $(1 \times 2^3) = 1 \times 8 = 8$
plus a count of "1" in the four column + $(1 \times 2^2) = 1 \times 4 = 4$
Plus a count of "0" in the two column + $(0 \times 2^1) = 0 \times 2 = 0$

Plus a count of "1" in the one column + $(1 \times 2^0) = 1 \times 1 = 1$
The summation of these counts = 13_{10}.

The octal system is base eight. This means there are eight discrete symbols to select from. The symbols are 0, 1, 2, 3, 4, 5, 6, and 7. These eight symbols are arranged in sequence to represent any numeric value. This works well as the count progresses from 0 to 7. But, when the count reaches 7, all available symbols are used. To progress the count, place weighting is required. A (0) is used to indicate the one's column is full and the count is resumed in the eight's column. Therefore any octal number is actually a summation of the counts in each column, with each column representing an additional multiple of eight.

Example: the number 243_8 represents,
A count of "2" in the sixty four column = $(2 \times 8^2) = 2 \times 64 = 128$
Plus count of "4" in the eight column + $(4 \times 8^1) = 4 \times 8 = 32$
Plus a count of "3" in the one column + $(3 \times 8^0) = 3 \times 1 = 3$
The summation of these counts is = 163_{10}.

The hexadecimal system is base sixteen. This means there are sixteen discrete symbols to select from. The symbols are 0, 1, 2, 3, 4, 5, 6, 7, 8, 9, A, B, C, D, E, and F. These sixteen symbols are arranged in sequence to represent any numeric value. This works well as the count progresses from 0 to F. But, when the count reaches F, all available symbols are used. To progress the count, place weighting is required. A (0) is used to indicate the one's column is full and the count is resumed in the sixteen's column. Therefore any hexadecimal number is actually a summation of the counts in each column, with each column representing an additional multiple of sixteen.

Example: the number $13C_{16}$ represents,
A count of "1" in the two hundred fifty six column = $(1 \times 16^2) = 1 \times 256 = 256$
Plus a count of "3" in the sixteen column + $(3 \times 16^1) = 3 \times 16 = 48$
Plus a count of "C" in the one column. + $(C \times 16^0) = C \times 1 = 12$
The summation of these counts is = 316_{10}.

Binary numbers are many times represented as groups of three or four. Three-bit binary numbers convert readily with octal, and four bit binary numbers convert readily with hexadecimal. It is important to begin the grouping with the least significant bit, 2^0. If the last grouping contains less than the three or four bits needed for conversion, zeros are placed in the empty high-order columns to complete the group.

Example: group this binary number into three bit binary format. 10111011_2
Begin with the lowest order bit and separate the bits into groups of three. $10 - 111 - 011_2$
The high-order group of three contains only two of the desired three bits.
Place a zero in the vacant highest order column $010 - 111 - 011_2$
The zero has no effect to the value of the term, but preserves the three bit-grouping formats.

Example: group this binary number into four bit binary format. 101011100_2
Begin with the lowest order bit and separate into groups of four. 1 - 0101 - 1100_2.
There are less than the required four bits in the higher order term.
Place zeros in the vacant high-order columns. 0001 0101 1100_2.
Again the zeros have no effect on the term, but merely preserve the four bit-grouping formats.

Three full place weights of binary total to a sum of seven. Starting the count at zero provides eight counts. Therefore, a three-bit binary number can represent any single octal symbol. Each additional digit in the octal number requires an additional set of three binary bits.

Example: $7_8 = 111_2$ and $17_8 = 001\ 111_2$.
Conversely, to convert any binary number into octal, group the binary digits in groups of three and convert each three-bit group into the octal equivalent.

Example: 101001111_2 separated into groups of three is $101\ 001\ 111_2$.
The highest order group converts to 5_8.
The next group converts to 1_8.
The low-order group converts to 7_8.
Therefore, the octal number is 517_8.

Four place weights of binary total to a sum of fifteen. Beginning the count with zero provides sixteen counts. Therefore, a four bit binary number can represent any single hexadecimal symbol. Each additional digit in the hexadecimal number requires an additional set of four binary bits.

Example: $F_{16} = 1111$ and $3C_{16} = 0011\ 1100$
Conversely, to convert any binary number into hexadecimal, group the binary digits into groups of four and convert each four-bit group into the hexadecimal equivalent.

Example: 101001111_2 separated into groups of four is $0001\ 0100\ 1111_2$. The highest order group converts to 1_{16}.
The next group converts to 4_{16}.
The low-order group converts to F_{16}.
Therefore, the hexadecimal number is $14F_{16}$.

8.3 Ladder Diagrams

Ladder diagrams are line code illustrations of programmable logic controller, process control software. Ladder diagrams document the manipulation and transmission of internal and external data. Rungs contain memory addresses, program code, or Boolean commands. The address for the device is a wiring block termination point, memory location, or a contact on a different rung. Unlike line drawings that only represent external contact data of actual

Mnemonic	Instruction	Mnemonic	Instruction
ADD	add	AVE	average
CMP	compare	CTD	count down
CTU	count up	DIV	divide
GEQ	greater than or equal to	LEQ	less than or equal to
MCR	master control reset	MUL	multiply
NEQ	not equal to	OTE	energize
OTL	latch	OTU	unlatch
PID	proportional/intigral/derivitive	RES	reset
TOF	timer off delay	TON	timer on delay

Figure 8.3-1 *Typical Mnemonics*

hardware components, ladder diagrams represent internal (software), and external (hardware), contact data. Memory addresses, program codes, and Boolean commands provide internal data. Field contacts from a pressure, temperature, level, flow, or a valve position switch provide external data. Contacts from pushbuttons, hand switches, relays, and timers also provide external data. Rung outputs drive alarms, stroke control valves, start/stop motors and pumps, transmit binary data, and provide logic states back to other circuits. Ladder diagrams are illustrated similar to line drawings, except the main lines (rails) on the ladder diagram are drawn vertically, and the individual circuits (rungs) are placed horizontally. The input/bus and output/bus are considered as rails of the ladder. Each subordinate circuit is an individual rung on the ladder. Each rung contains a condition instruction (input) and a command instruction (output). Each successive rung on the ladder is sequentially acknowledged. As the condition instruction is realized (made true), the command instruction on that rung is executed. The functional description and address identify symbols on each rung. The type of instruction is identified as I (input), O (output), T (times), or C (counter). Figure 8.3-1 lists the actual instruction as identified by a mnemonic symbol assigned by the manufacturer.

8.4 Input/Output Addressing

Instruction addresses are presented in binary, octal, or hexadecimal format. The first set of numbers identifies the physical rack. The second set identifies the physical module, and the third set identifies the actual (external) or virtual (internal) terminal. The actual address for the device is a wiring block termination point. Figure 8.4-1 depicts the virtual address memory mapped to the specific data bit in the word address representing the particular hardware module. The octal number IO:00/07/12 locates the termination point in rack (cabinet) 00, input module (card) 07, and terminal (screw) 12, shown in Figure 8.4-2. This data point is mirrored in the memory map at word IO:00/07/12. The ladder diagram in Figure 8.4-3 depicts logic to prevent the low oil pressure alarm from coming on until the oil pressure is low for longer than sixty seconds. This type of logic is also used to prevent nuisance alarms from noise, spikes, and transient conditions.

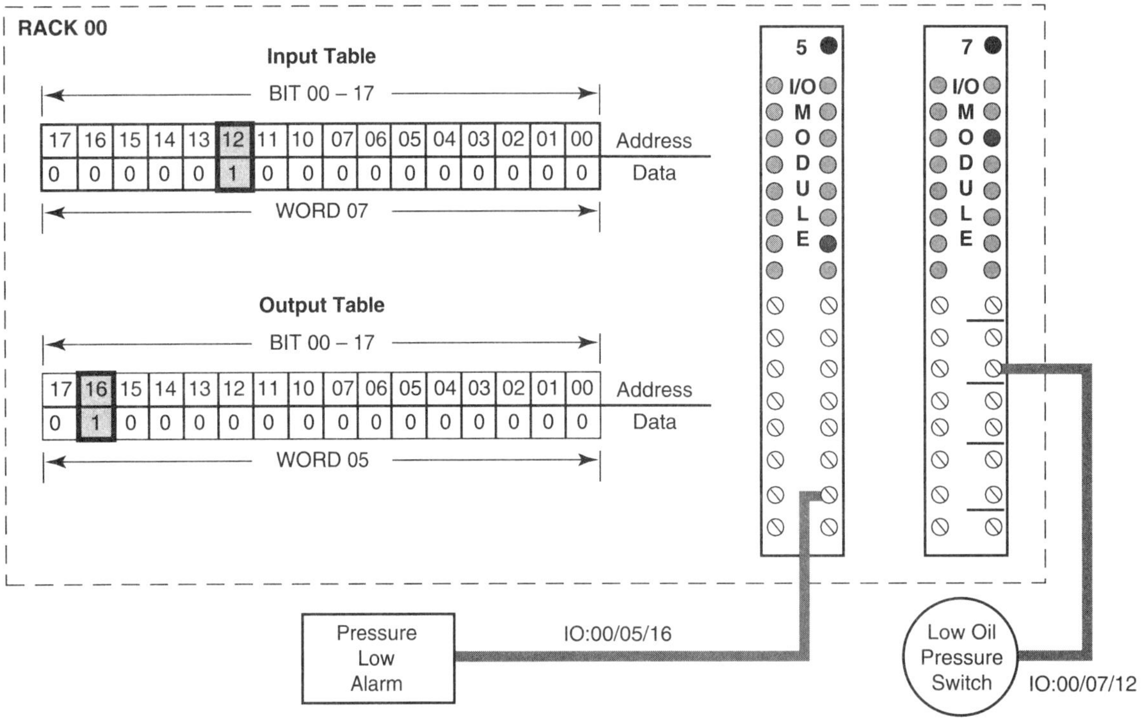

Figure 8.4-1 *Internal I/O Addressing*

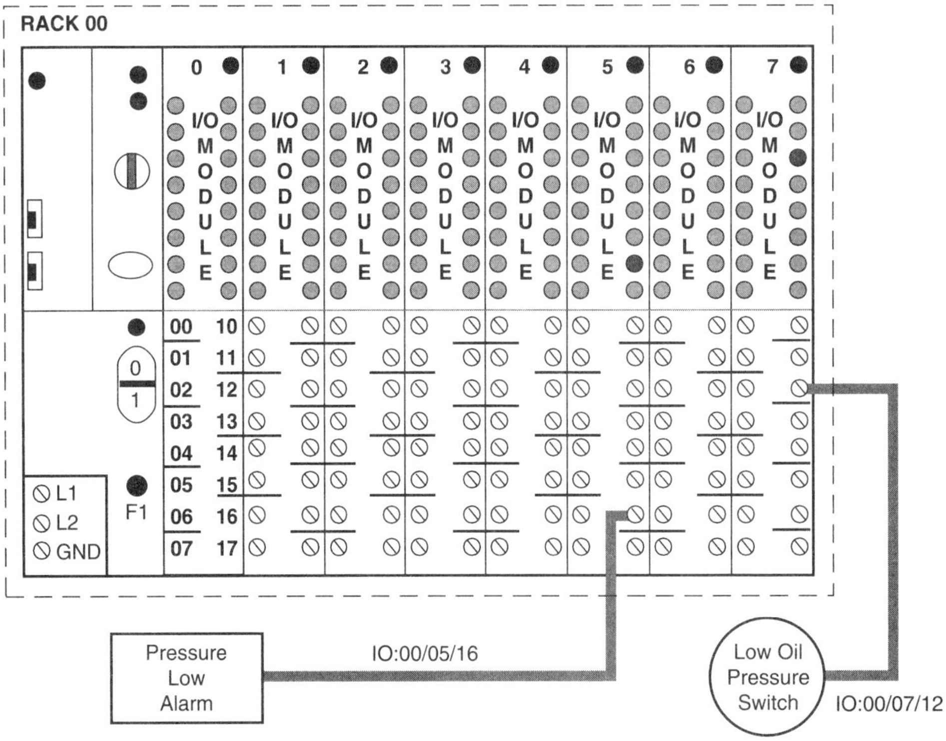

Figure 8.4-2 *External I/O Addressing*

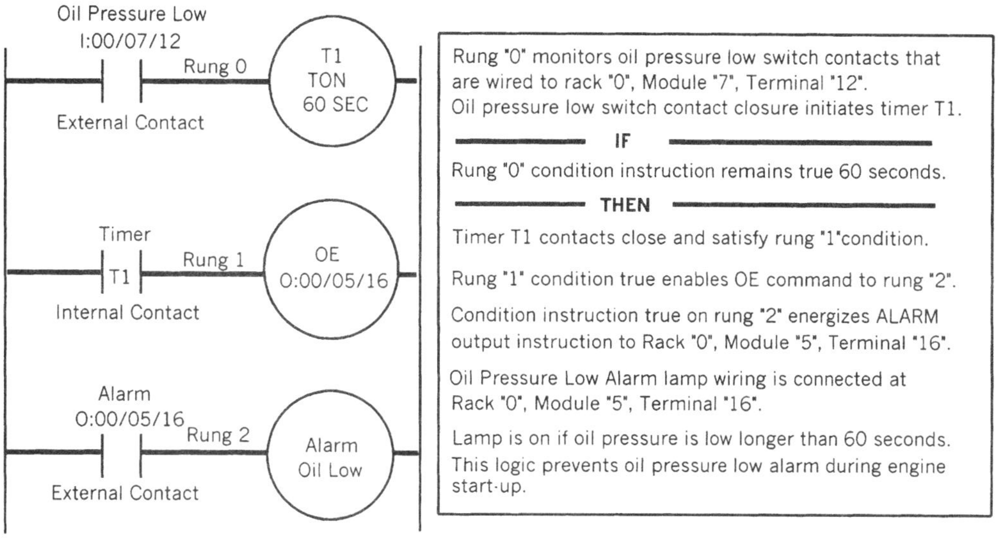

Figure 8.4-3 *Ladder Logic Diagram*

Figure 8.4-4 *Programmable Logic Control Rack*

8.5 Troubleshooting

Termination point status lights are provided on each binary I/O module. If the condition at any particular termination point is true, the light illuminates. If the condition monitored at the termination point violates the program sequence, a discordance alarm is generated, and the status light blinks until the discordance is eliminated. A DVM can be used to monitor the input/output terminals and verify the LED status indication is correct. Exercise caution and verify the DVM is configured to monitor the value expected. The module type indicates the expected value. Typical values may be 12–120 VAC, 12–120 VDC, or 4–20 mADC. Internal fuses at each I/O termination point protect the processor from excess current. The I/O modules have a single blown fuse indicator that illuminates if any fuse in the module fails to conduct. Fuse replacement may require partial removal of the affected module. Never remove or install an I/O module until the rack is de-energized. The interface and processor modules also display current internal status update data. They each provide several LEDs to indicate system health and activity. A local communication port is often provided on the processor. A handheld communicator or laptop computer containing the appropriate software can be used to read input status, read output status, or toggle an output point. A section of memory is usually dedicated to store error flags or error bits. The communicator can be used to view this information. The processor module is provided with a Read/Write/Run key switch to prevent accidental or unauthorized changes to the software. Most PLC related failures are actually failed field devices providing erroneous information, or failing to actuate on demand.

Chapter Eight Summary:

- Programmable logic controllers are microprocessor based sequential step software configurable controllers.
- The hardware consists of a mounting rack, a power supply, a communication interface module, a processor module, and various application specific input/output modules.
- The mounting rack provides a stable platform for the modules, and contains the back-plane bussing network that distributes power and data through the rack.
- The power supply provides back-plane power to the modules and signal power or contact wetting voltage for the output modules.
- The communication interface module provides data transport links to other racks in the system, and RS232 serial or RJ11 phone connections to facilitate user programming with a laptop computer or hand held communicator.
- The processor module (CPU) provides memory locations for executive programs, application programs (ladder logic), data bits, and registers. It also performs all logic functions, and executes the application program.
- Input/output modules receive and transmit data to and from field devices, provide electrical isolation to protect the CPU, and transpose data into the appropriate retransmission format.
- The type of data transmitted determines input/output module selection. Typical input modules are AC volts binary input, DC volts binary input, DC volts analog input, DC milliamps analog input, thermocouple analog input, and RTD analog input. Typical output modules are AC volt binary, DC volt binary, and DC volts analog.
- Alternate numbering systems used are binary (base two), octal (base eight), and hexadecimal (base sixteen).
- The binary system is base two, there are only two symbols to select from 0 and 1.
- The octal system is base eight, there are eight discrete symbols to select from 0, 1, 2, 3, 4, 5, 6, and 7.
- The hexadecimal system is base sixteen, there are sixteen discrete symbols to select from; 0, 1, 2, 3, 4, 5, 6, 7, 8, 9, A, B, C, D, E, and F.
- Ladder diagrams are line illustrations of process control application software.
- Ladder diagrams are represented as two parallel vertical lines connected by horizontal rungs. Rungs depict inputs from and outputs to field devices, including the specific address of each data point.
- The functional description and address identify symbols on each rung. The type of instruction is identified as I (input), O (output), T (times), or C (counter). The actual instruction is identified by a mnemonic symbol assigned by the manufacturer.
- Each rung contains a condition instruction (input) and a command instruction (output). As each successive rung on the ladder is sequentially acknowledged, the condition instruction is realized (made true), and the command instruction (mnemonic) on that rung is executed.

- Contacts from pushbuttons, hand switches, relays, timers, pressure, temperature, level, flow, or valve position switches provide external data.
- Rung outputs drive alarms, stroke control valves, start/stop motors and pumps, transmit binary data, and provide logic states back to other circuits.
- The address of the instruction is presented in binary, octal, or hexadecimal format.
- The first set of numbers identifies the physical rack. The second set identifies the physical module, and the third set identifies the actual (external) terminal. The actual address for the device is a wiring block termination point. The virtual address is memory mapped to the specific data bit in the word address representing the particular rack and module.
- Termination point status lights are provided on each binary I/O module. If the condition at any particular termination point is true, the light illuminates.
- If the condition monitored at the termination point violates the program sequence, a discordance alarm is generated, and the status light blinks until the discordance is eliminated.
- Internal fuses at each I/O termination point protect the processor from excess current. The I/O modules have a single blown fuse indicator that illuminates if any fuse in the module fails to conduct.
- The interface and processor modules also display internal status update data. They each provide several LEDs to indicate system health and activity. The processor module is provided with a Read/Write/Run key switch to prevent accidental or unauthorized changes to the software.

Chapter Eight Review Questions:

1. State the purpose and function of a programmable logic controller (PLC).
2. List the five fundamental hardware components of a PLC.
3. Describe the function of each individual hardware component of a PLC.
4. List the typical inputs/outputs monitored by a PLC.
5. Convert 001 110 binary into octal and hexadecimal equivalents.
6. Draw ladder logic symbols for a normal open contact and a normal closed contact.
7. List typical logic operands used by a PLC.
8. Draw a simple switch alarm ladder logic diagram.
9. Locate the termination point represented by input table word IO:00/05/bit12.
10. State the method to determine if the anticipated field input is recognized by the PLC.
11. State the method to determine if a module contains a faulty fuse.
12. State the method to determine if the interface module is communicating.
13. State the method to determine if the power supply is O.K.
14. State the location of the self-health status monitoring LEDs.
15. State the conditions required to generate a discordance alarm.

Chapter 9

Distributive Control Systems

OBJECTIVES

Upon completion of Chapter Nine the student will be able to:

- *State the definition of a distributive control system (DCS).*
- *List the three different technologies that comprise a distributive control system.*
- *List the basic hardware components used in a distributive control system.*
- *State the purpose of the master control unit.*
- *State the purpose of field mounted intelligent instrumentation devices.*
- *State the purpose of an operator interface (terminal).*
- *State the definition of a bus as it relates to distributive control.*
- *State the purpose of the trunk line as it relates to distributive control.*
- *State the purpose of the bus as it relates to distributive control.*
- *State the definition of a node as it relates to distributive control.*
- *List several types of conductors used in distributive control busses.*
- *Identify components used to physically connect devices to the bus.*
- *State the purpose of individual remote power*

supplies on the bus.
- *State the purpose of terminators as it relates to distributive control.*
- *State the purpose of amplifiers and repeaters as it relates to distributive control.*
- *Describe the function of amplifiers as it relates to distributive control.*
- *Describe the function of repeaters as it relates to distributive control.*
- *State the purpose of interface devices as it relates to distributive control.*
- *List several interface devices.*
- *State the definition of protocol.*
- *State the function of protocol.*
- *State the purpose of protocol.*
- *List the information included in instrument tag data as it relates to distributive control.*
- *State the purpose of tag data as it relates to distributive control.*
- *State the primary functions of application software.*
- *State three methods of human interface with the bus.*
- *List several methods used to display process data by the application software.*
- *State the purpose of the polling process as it relates to distributive control.*
- *State the definition of system topology as it relates to distributive control.*
- *State the definition of a network as it relates to distributive control.*
- *List three common topologies.*
- *Develop a simple diagram depicting star topology.*
- *Develop a simple diagram depicting multi-drop topology.*
- *Develop a simple diagram depicting ring topology.*
- *Describe a data collision.*
- *Explain the relationship between traffic volume and bus capacity.*
- *Explain the relationship between line length, signal strength, and noise.*
- *Differentiate between introduced noise and induced noise.*
- *List the most common problems to the bus.*

9.0 Introduction

Perhaps the most significant technological development in modern instrumentation is the advent of distributive control systems. A distributive control system is an interactive network of instrumentation, communication, and control devices connected by a common communication link. Distributive control systems are actually a combination of three different technologies; intelligent (microprocessor based) instruments, communication networks (data links), and application software.

9.1 Hardware

Distributive control system hardware consists of: a master control device such as a PLC or Host CPU, an array of intelligent monitoring and control instruments (transmitters, controllers, and valves), various peripheral devices, such as a remote operator interface CPU, display, and keypad, Figure 9.1-1, and the cabling and communication interface components that create the data link.

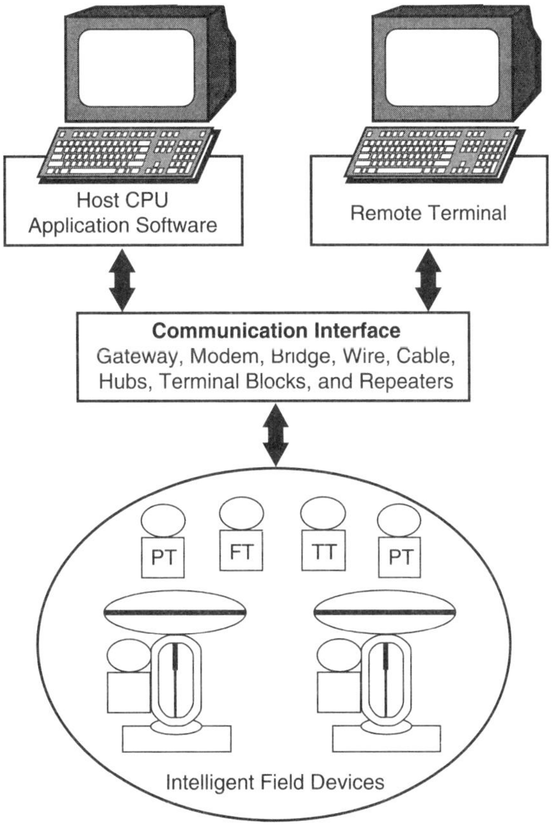

Figure 9.1-1 *Distributive Control System*

The master control unit contains the display and control (host application) software and regulates network communication. It offers the highest level of user interaction. Program uploads and downloads, configuration changes, and actual active process commands are generally only authorized at the host CPU. Intelligent instrumentation devices such as transmitters, controllers, and control valves are mounted in the field to measure and control the process variables. An operator interface (terminal) that consists of a processor, a display screen, and a keypad allows the operator to view current process and equipment condition, request historical data, initiate trending of particular points, and verify field device status.

A bus is any communication interface consists of cabling, terminations, power supplies, and necessary interface modules. The main bus (data hi-way) is often called the trunk line. Busses connect network devices together and permit the transmission and reception of data. Each addressed device on the bus is a node. Conductors are usually twisted pair shielded wire, coaxial cable, or multi-conductor cable. Special T-connectors and node modules physically connect devices to the bus. Individual remote power supplies are often installed to provide regulated operating voltage to the instruments on each section of the bus. Terminators are installed at the end of each drop to stabilize the signal and reduce line noise. Terminators are resistor/inductor circuits tuned to match the cable impedance. Amplifiers and repeaters are installed at critical points along the bus to maintain signal integrity and increase the useful length of the runs. Amplifiers boost the incoming signal and send it further down the bus. Amplifiers have negligible effect on transmission speed, but any distortion or noise present in the original signal is also amplified and retransmitted. Repeaters sample the original incoming signal and regenerate a duplicate signal for retransmission. Signal quality is maintained, but transmission speed is affected. Interface devices are installed to connect devices to bus segments and facilitate cross protocol communication between busses. Interface devices are gateways, bridges, and scanners.

9.2 Software

Protocol is a set of rules that govern the composition and transmission of binary data. Binary communication is transmitted as a string of ones and zeros. Protocol functions to segregate the ones and zeros into specific ordered groups or frames of serial data that are readily interpreted by devices on the bus. Serial data strings represent a complete and intelligent statement or command. Synchronized clock pulses establish a common point of reference for the binary data. Each serial transmission contains a designated start and stop code. Addresses, commands, and data are each assigned a specific ordered sequence within the message, and the type of error checking is identified.

Intelligent instruments contain an independent programmable processor with memory, and posse's self-diagnostic and communication capability. On-board processors contain a packet of information specific to each individual instrument. This information is called tag data. Tag data includes the polling address, manufactures specification, calibration data, and tun-

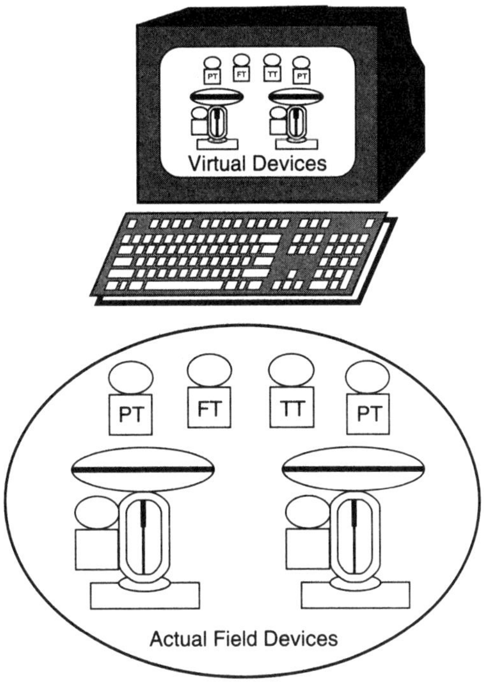

Figure 9.2-1 *Application Software*

ing algorithms unique to the particular device. Application software uses tag data polling addresses to identify and verify each particular instrument. Process variable updates as well as control and configure command communications are relegated to a particular address.

Handheld remote communicators can interface with the distributive control system at any point on the bus. Communicators use instrument tag data to identify and communicate with all intelligent instruments on the bus. The full array of applications, however, is not usually accessible with the handheld.

Application software performs several primary functions. It administers and regulates bus communication, data transfer, and storage. It monitors and controls process variables. It implements self-health and diagnostic analysis, and supports human machine interface. Application software contains a "virtual" process system that mimics the actual hardware. The control software utilizes instrument addresses to send and receive data from each instrument on the data link. As actual field devices provide new information, the software updates the virtual representations of the device, Figure 9.2-1. Application software collects, analyzes, and updates data from the monitored points. It generates graphic displays, trends, plots, and graphs of process variable values, and automatically processes actuations, notifications, and alarms. The program also monitors overall system health. It initiates a polling process that performs routine instrument operability checks and verifies communication integrity of the bus. Software also supports operator interface capability. Additional application software functions available through the host or remote terminal include; data log recovery/display, cross calibration comparisons, first out event analysis, flow chart programming, troubleshooting, corrective action statements, diagnostic analysis, and predictive maintenance schedules.

9.3 Topology

Distributive control systems are configured with various topologies. System topology refers to the particular technique employed to physically connect individual devices throughout the network. A network is a group of field and control devices connected through a common medium. Each addressed device connected to the network is a node. Common topologies are star, multi-drop, and ring. Star topology connects several point-to-point devices to a central communicator, Figure 9.3-1. Multi-drop topology connects all nodes in parallel to a single cable, Figure 9.3-2. Ring topology electrically connects the nodes in a series loop. Ring topology uses each device on the ring as a transceiver, repeater, Figure 9.3-3. Communication passes through all devices, and only the specific devices addressed respond to the command. Combination topology is quite common, and many times results as additional devices and controls are incorporated into the original design, Figure 9.3-4.

Many distributive control malfunctions are self-diagnosed and self-corrected. If a particular instrument or node experiences a failure, the master controller sets an alert flag and removes that device or group of devices from service. If parallel channels are available, the host transfers local control to them. Certain programs substitute an averaged signal in place of missing process data until the failure is corrected. Most distributive control network

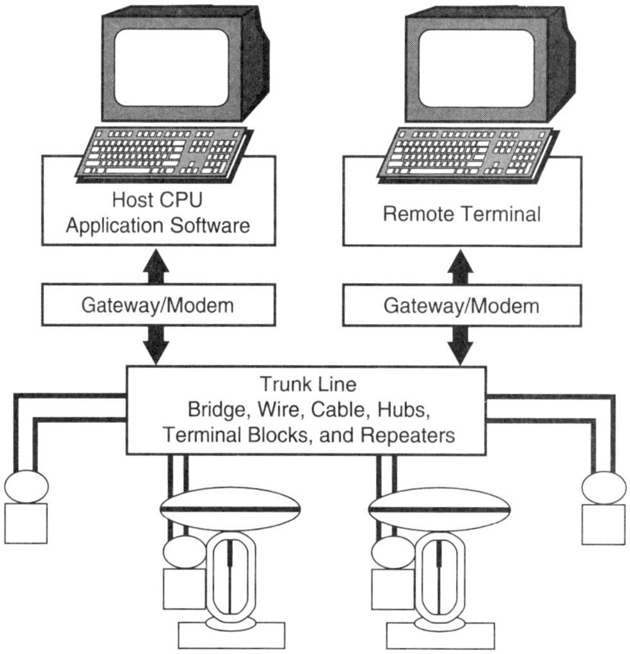

Figure 9.3-1 *Star Topology*

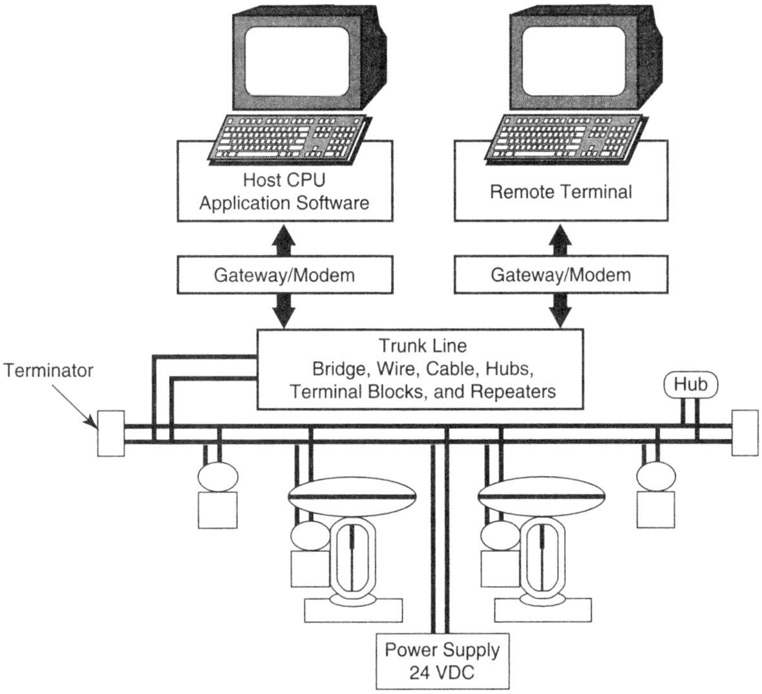

Figure 9.3-2 *Multi-Drop Topology*

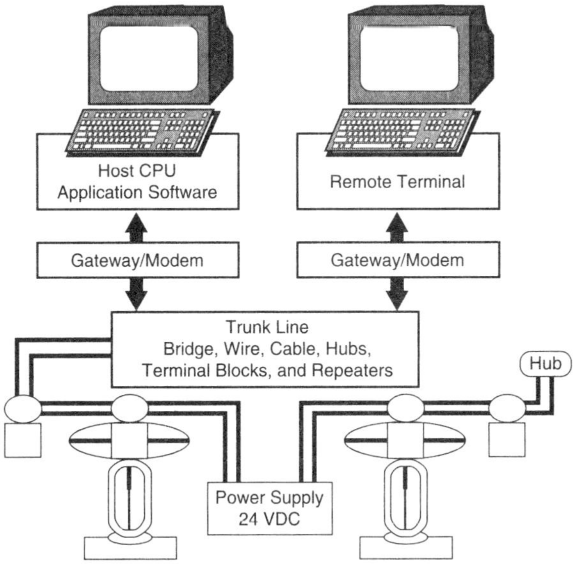

Figure 9.3-3 *Ring Topology*

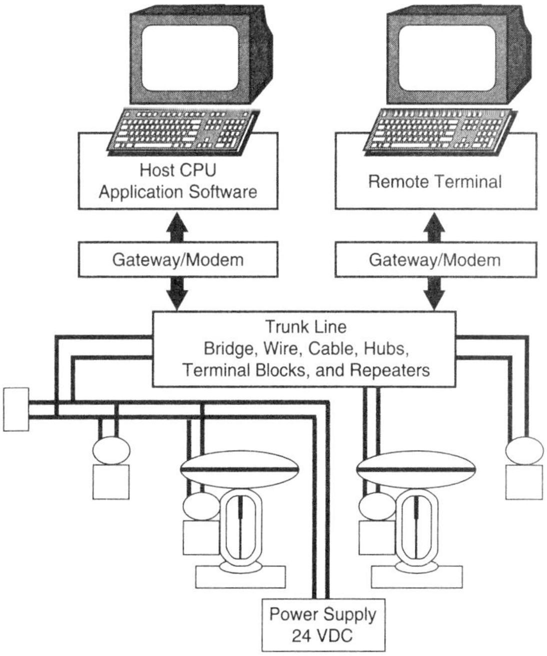

Figure 9.3-4 *Combined Topology*

problems appear shortly after the system is placed in service. Data collisions occur when several different devices attempt to transmit simultaneously. When several different devices talk at once it is difficult for the devices that are listening to make much sense of it all. High traffic evolutions such as system start-up can stress a system that is already loaded to capacity and induce problems that are not evident under normal conditions. Excessive line length increases noise and attenuates the data signal. Grounding problems, power supply failures, and noise are the most common problems to the bus.

9.4 Troubleshooting

There are several techniques to determine the integrity of bus cabling. They do however, require specific skill and knowledge. Incorrect use of measuring and test equipment can result in personal injury and additional damage to bus devices. Only trained personnel should perform the following system checks. All measuring and test equipment connected to the bus must be isolated from plant ground. Use battery powered devices, or ensure the proper isolating transformer is used for AC powered equipment. Always determine and maintain correct polarity. A DVM connected to measure ohms or amperes initiates a direct short, current path, through the meter. To verify correct cabling, de-energize and isolate the bus. To isolate a short circuit condition, measure the resistance of each conductor to ground with a digital voltmeter. A high resistance indicates an open condition, and a low resistance indicates a short. Power and signal conductors should read practically infinite resistance to ground. The shield and drain conductor should read practically zero resistance to ground when terminated, and infinity when the single point ground is lifted. All conductors should read infinite resistance from one to the other, and infinite resistance to the shield. To verify

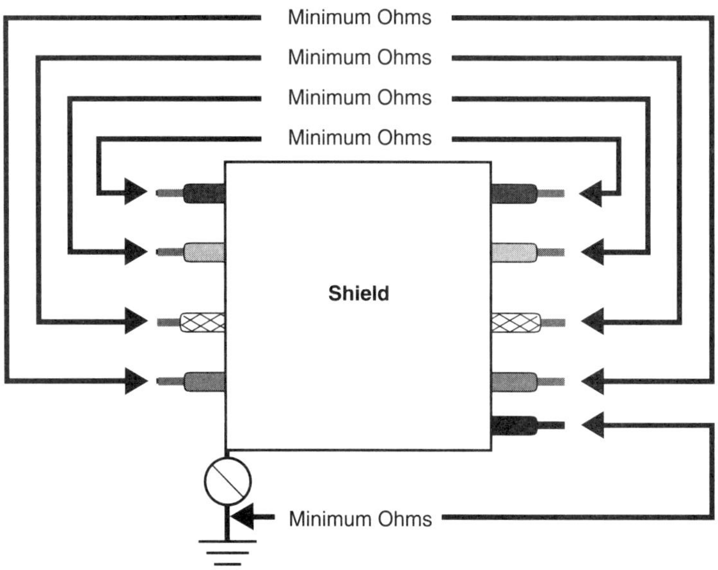

Figure 9.4-1 *Ideal Conductors/Minimum Resistance*

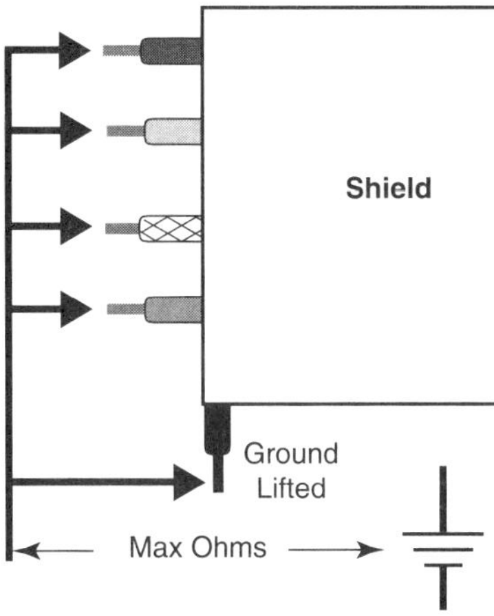

Figure 9.4-2 *Floated Conductors Isolated*

continuity of the conductors, sequentially and temporarily tie each conductor to ground at one end of the run, and measure conductor resistance to ground from the other end of the run. A good conductor should read extremely low ohms. High ohms resistance indicates a damaged or cross-wired conductor. Disconnect the temporary ground before proceeding to the next conductor. Continue until all conductors are verified, Figures 9.4-1 and 9.4-2.

Power supply problems include low voltage, low current, and noise. To investigate the power supply, disconnect the load and measure the output voltage with a DVM. If the voltage is incorrect, replace the power supply. If the voltage is within specified tolerances, monitor current and voltage as the load is reintroduced. Low voltage and/or high current indicate an excessive load, but not necessarily a ground fault. Compare the current requirements of the combined devices on the bus to the power supply specifications. If possible, measure voltage at several points along the bus and attempt to locate any excessive voltage drops. Noise can reduce signal integrity; especially where line length is excessive. Devices on the bus, particularly power supplies, can introduce noise. High-energy electromagnetic fields can induce noise on to the bus. To determine noise on a DC power supply, measure for milli-volts AC at the power supply output, and compare the measurement to manufactures' specifications. Induced noise often occurs as an intermittent spike, and is difficult to document, even with recorders and oscilloscopes. Visually inspect the cable runs for proximity to high-energy devices or activities. High voltage motor controls, welding, and radiography can induce noise on the bus. Re-route the cable or increase cable shielding in these areas. Cycle suspected equipment and attempt to correlate data transmission faults to the operation of specific equipment.

Chapter Nine Summary:

- A distributive control system is an interactive network of instrumentation, communication, and control devices connected by a common communication link.
- Distributive control systems are actually a combination of three different technologies; intelligent (microprocessor based) instruments, communication networks (data links), and application software.
- Distributive control system hardware consists of: a master control device such as a PLC or host CPU, an array of intelligent monitoring and control instruments (transmitters, controllers, and valves), various peripheral devices, such as a remote operator interface CPU, display, and keypad, and the cabling and communication interface components that create the data link.
- The master control unit contains the display and control (host application) software and regulates network communication.
- Intelligent instrumentation devices such as transmitters, controllers, and control valves are mounted in the field to measure and control the process variables.
- An operator interface (terminal) that consists of a processor, a display screen, and a keypad allows the operator to view current process and equipment condition, request historical data, initiate trending of particular points, and verify field device status.
- A bus is any communication interface consists of cabling, terminations, power supplies, and necessary interface modules.
- The main bus, data hi-way, is often called the trunk line.
- Busses connect network devices together and permit the transmission and reception of data.
- Each addressed device on the bus is a node.
- Conductors are usually a twisted pair shielded wire, coaxial cable, or multi-conductor cable.
- Special T-connectors and node modules physically connect devices to the bus.
- Individual remote power supplies are often installed to provide regulated operating voltage to the instruments on each section of the bus.
- Terminators are installed at the end of each drop to stabilize the signal and reduce line noise.
- Amplifiers and repeaters are installed at critical points along the bus to maintain signal integrity and increase the useful length of the runs.
- Amplifiers boost the incoming signal and send it further down the bus.
- Repeaters sample the original incoming signal and regenerate a duplicate signal for retransmission.
- Interface devices are installed to connect devices to bus segments and facilitates cross protocol communication between busses.

- Interface devices are gateways, bridges, and scanners.
- Protocol is a set of rules that governs the composition and transmission of binary data. Protocol segregates the ones and zeros into specific ordered groups or frames of serial data that is readily interpreted by devices on the bus.
- Handheld remote communicators interface with the distributive control system from any point on the bus.
- Communicators use instrument tag data to identify and communicate with all intelligent instruments on the bus.
- Application software performs several primary functions. It administers and regulates bus communication, data transfer, and storage. It monitors and controls process variables. It implements self-health and diagnostic analysis, and supports human machine interface.
- System topology refers to the technique employed to physically or electrically connect individual devices throughout the network.
- A network is a group of field and control devices connected through a common medium.
- Common topologies are star, multi-drop, and ring.
- Star topology connects several point-to-point devices to a central communicator.
- Multi-drop topology connects all nodes in parallel to a single cable.
- Ring topology electrically connects the nodes in a series loop.
- Data collisions occur when several different devices attempt to transmit simultaneously.
- High traffic evolutions such as system start-up can stress a system that is already loaded to capacity and induce problems that are not evident under normal conditions.
- Grounding problems, power supply failures, and noise are the most common problems to the bus.

Chapter Nine Review Questions:

1. State the definition of a distributive control system.
2. List the three different technologies that comprise a distributive control system.
3. List the basic hardware components used in a distributive control system.
4. State the purpose of the master control unit.
5. State the purpose of field mounted intelligent instrumentation devices.
6. State the purpose of an operator interface (terminal).
7. State the definition of a bus as it relates to distributive control.
8. State the purpose of the trunk line as it relates to distributive control.
9. State the purpose of the bus as it relates to distributive control.
10. State the definition of a node as it relates to distributive control.
11. List several types of conductors used in distributive control busses.
12. Identify components used to physically connect devices to the bus.
13. State the purpose of individual remote power supplies on the bus.
14. State the purpose of terminators as it relates to distributive control.
15. State the purpose of amplifiers and repeaters as it relates to distributive control.
16. Describe the function of amplifiers as it relates to distributive control.
17. Describe the function of repeaters as it relates to distributive control.
18. State the purpose of interface devices as it relates to distributive control.
19. List several interface devices.
20. State the definition of protocol.
21. State the function of protocol.
22. State the purpose of protocol.
23. List the information included in instrument tag data as it relates to distributive control.
24. State the purpose of tag data as it relates to distributive control.
25. State the primary functions of application software.
26. State three methods of human interface with the bus.
27. List several methods used to display process data by the application software.
28. State the purpose of the polling process as it relates to distributive control.
29. State the definition of system topology as it relates to distributive control.
30. State the definition of a network as it relates to distributive control.
31. List three common topologies.
32. Develop a simple diagram depicting star topology.
33. Develop a simple diagram depicting multi-drop topology.
34. Develop a simple diagram depicting ring topology.
35. Describe a data collision.
36. Explain the relationship between traffic volume and bus capacity.
37. Explain the relationship between line length, signal strength, and noise.
38. Differentiate between introduced noise and induced noise.
39. List three of the most common problems to the bus.

Chapter 10

Practical Application of Knowledge

OBJECTIVES

Upon completion of Chapter Ten the student will be able to:

- *List steps required prior to performing any equipment manipulations.*
- *State the common method of mounting tank level instruments.*
- *State the adverse effects of manipulating tank level instruments mounted on a common standpipe.*
- *State the proper valve sequences to remove and restore level instruments to service.*
- *State the reason to use caution when venting or draining a level instrument.*
- *State the adverse effects of manipulating instruments that share a common sensing line.*
- *State two sensing line conditions that create significant measurement error.*
- *State a condition to avoid when manipulating differential pressure isolation valves.*
- *State the normal valve sequences to remove and restore a transmitter to service.*
- *State the proper sequence to execute an auto/manual transfer.*
- *State the proper sequence to execute a manual/auto transfer.*

- *State the proper sequence to remove a valve from service.*
- *State the proper sequence to restore a valve to service in auto-control.*
- *State the proper sequence to restore a valve to service in manual-control.*
- *State the three common types of control response.*
- *Describe quarter wave dampening (ultimate period).*
- *Describe minimum disturbance tuning.*
- *Describe minimum area tuning.*
- *Describe the relationship between reset output and valve position.*
- *Describe the relationship between derivative output and valve position.*
- *Describe the relationship between proportional output and valve position.*
- *Explain the dynamic control response of proportional (P).*
- *Explain the dynamic control response of reset (I).*
- *Explain the dynamic control response of rate (D).*

10.0 Introduction

The following information is represented as a fundamental approach or guide only. Only trained and qualified technicians with extensive hands-on training for each particular loop should attempt field tuning or manipulating active process systems and control loops. Before any equipment manipulations, communications must be established between the board and field operators to review all affected control schemes, develop guidelines, and establish monitoring of the process variable.

10.1 Removing and Restoring Instruments to Service

Tank level instruments are often connected to a common standpipe as in Figure 10.1-1. Any change or disturbance to the standpipe inlet/outlet valves will affect all instruments attached to the common standpipe. Manipulating level switch and level control isolation valves can unexpectedly change the stand-pipe level and result in unanticipated alarms and actuations. Level instruments typically have a high side inlet valve and a vent valve installed on the upper inlet, and a low side isolation valve and drain valve connected to the lower inlet. Fully close both isolation valves before opening the vent or drain valves. It is critical to remember that even with both upper and lower isolation valves closed, full process pressure is still trapped in the pot. Use extreme caution when opening the vent and/or drain.

Transmitters are often removed and restored to service. Although it is poor practice, instruments may share common sensing lines. If several components are connected, as in

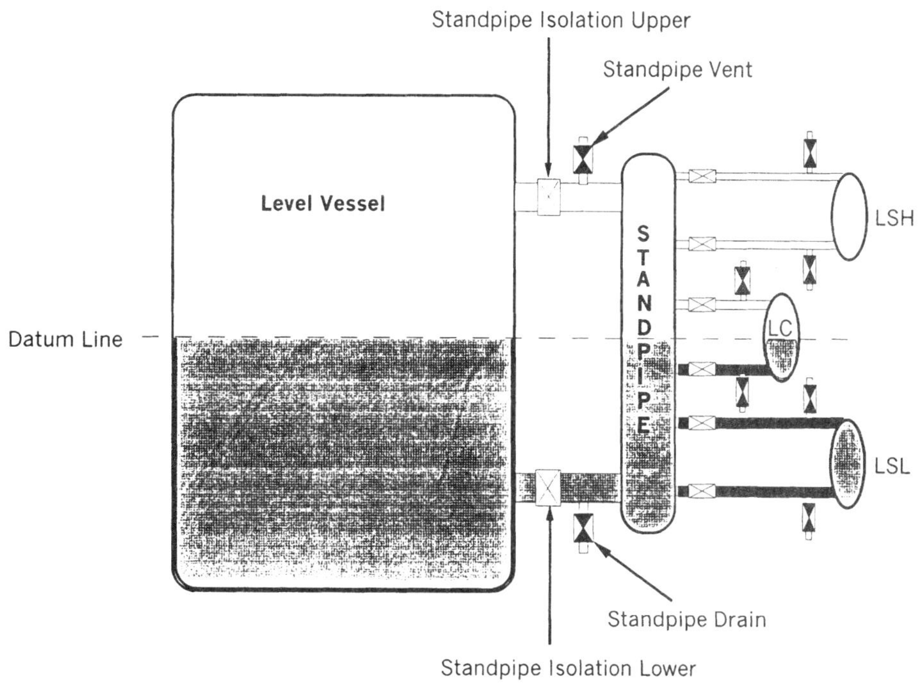

Figure 10.1-1 *Standpipe and Instrument Isolations*

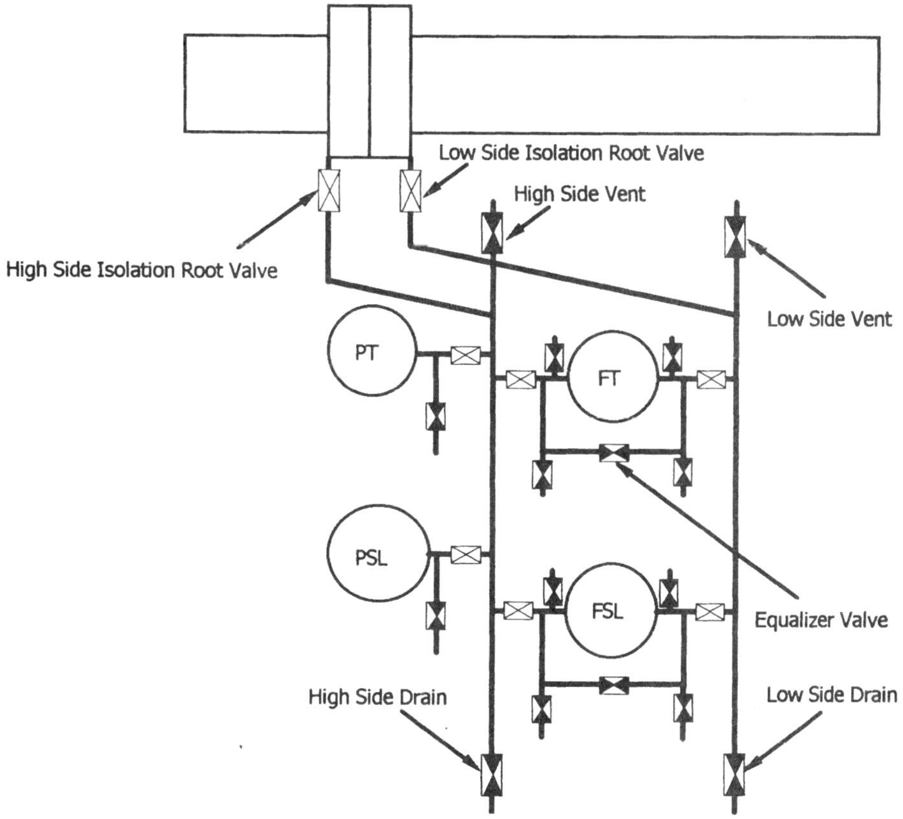

Figure 10.1-2 *Common Sensing Line Isolations*

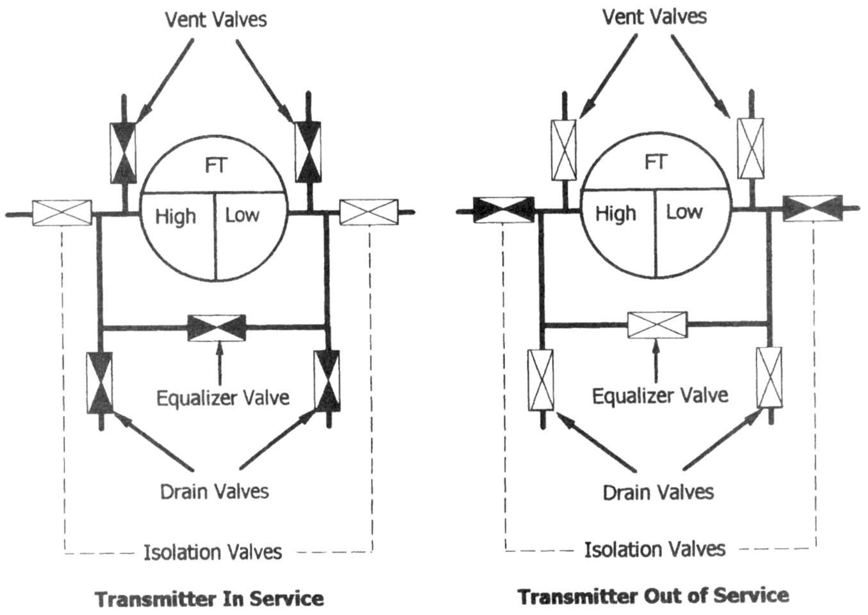

Figure 10.1-3 *Transmitter Isolation Valves*

Figure 10.1-2, special precautions are required to prevent upsets. Manipulating transmitter isolation valves can induce spikes on the sensing lines, which could result in an unanticipated alarm or actuation. Many instruments are installed with filled sensing lines. Flow transmitters are extremely sensitive to sense line conditions. They require the sense line to be completely dry and devoid of all moisture, or completely solid and devoid of all air. A loss of sensing line reference fluid, or entry of air into sealed systems, creates significant measurement error. Long runs of sensing line are installed at a slope with vent and drain valves installed to facilitate venting and/or filling. To prevent damage to the sensor the transmitter is never exposed to excessive differential pressure. The traditional method to isolate a transmitter is to first close the high side process inlet valve, refer to Figure 10.1-3. Then open the transmitter equalization valve. This ensures that the low-side inlet pressure is applied equally to both sides of the transmitter. Once the transmitter is equalized, close the low-side inlet valve. Transmitters are restored to service in a similar manner. First, open the transmitter equalization valve. Then, open the low-side inlet valve. Allow sufficient time for the transmitter to equalize, and close the transmitter equalization valve. Open the transmitter high-side inlet valve and restore the transmitter to service.

10.2 Auto / Manual Transfer

Auto/manual transfer is often required to support maintenance activities and other evolutions. Bumpless transfer is the ability to switch between automatic and manual control operation without disturbing the process or repositioning the final control element. It is critical to execute a bumpless transfer, ensure the transfer is transparent and that the final control element remains stable. While in auto-control, adjust the manual output to match the auto output. Once the outputs are equal, transfer to manual-control and ramp the final element to the desired position. Adjustments are identified in Figure 10.2-1.

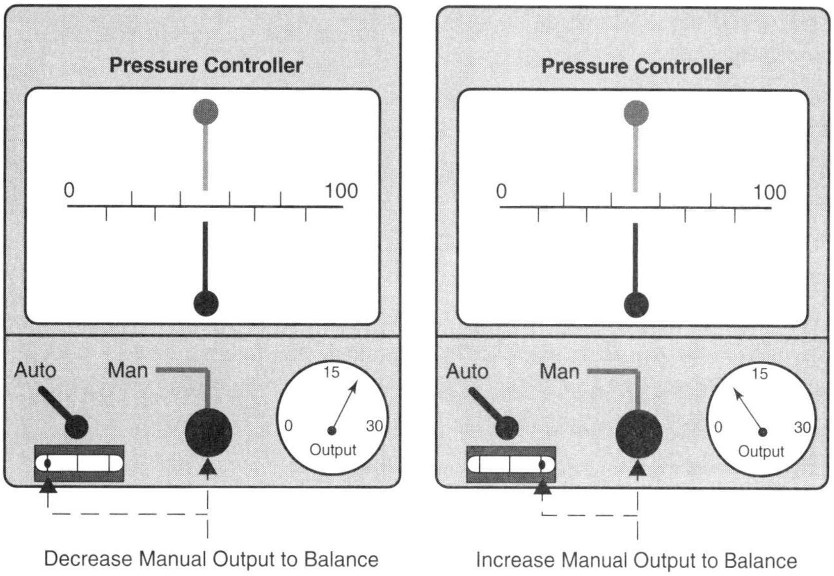

Figure 10.2-1 *Auto/Manual Transfer*

Manual/auto transfer is accomplished much the same way as auto/manual transfer. Manipulate the controller set point adjustment to match the auto-output signal to the manual demand signal. Once the outputs are equal, place system in auto-control and adjust the controller set-point as required to ramp the final control element to the desired position. Adjustments are identified in Figure 10.2-2. In any event, always ensure that the automatic demand signal and the manual demand signal are equal before attempting the transfer.

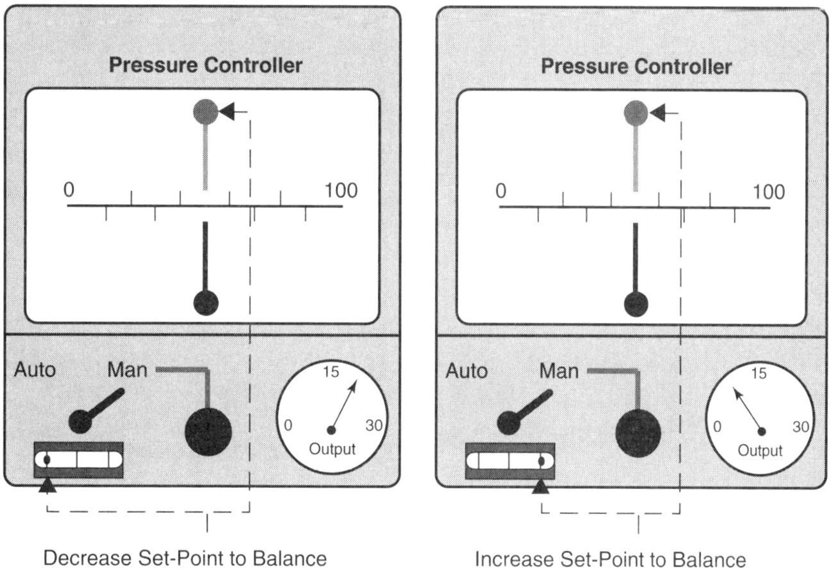

Figure 10.2-2 *Manual/Auto Transfer*

10.3 Removing and Restoring Valves to Service

The preferred method for removing a valve from service is to maintain auto-control and slowly throttle the by-pass valve into service. Observe the primary final control element move toward full closed as the by-pass valve assumes the flow. Once the primary valve reaches full closed, manipulate the controller set-point to overdrive the primary valve (hard seat). Set the manual-control output to match the auto-control output and transfer to manual-control. Secure the primary final-control element. This sequence is depicted in Figures 10.3-1, 10.3-2, and 10.3-3.

There are also different methods for restoring a valve to service. The preferred method is with the controller in automatic control. First manipulate the controller set-point to match the auto-output signal to the manual demand signal. Allow sufficient time, as auto-response may be delayed if the controller is in a reset wind-up condition. Once the demand signals match, place the controller in automatic. Slowly throttle the by-pass valve closed, and observe the primary final control element modulate open and into control as the by-pass valve assumes less flow. Adjust the controller set-point to ramp the primary valve to the desired control position.

To restore a valve to service with the controller in manual-control, ensure the controller manual demand signal matches the primary valve position. Slowly adjust the manual output to ramp the primary valve toward the desired position. Throttle the by-pass closed, and verify that the controller auto-signal responds to process conditions. Allow sufficient time, as auto-response may be delayed if the controller is in a reset wind-up condition. Set the manual-control output to match the auto-control output and place the controller in auto-control. Adjust the controller set-point to drive the primary valve to the desired position.

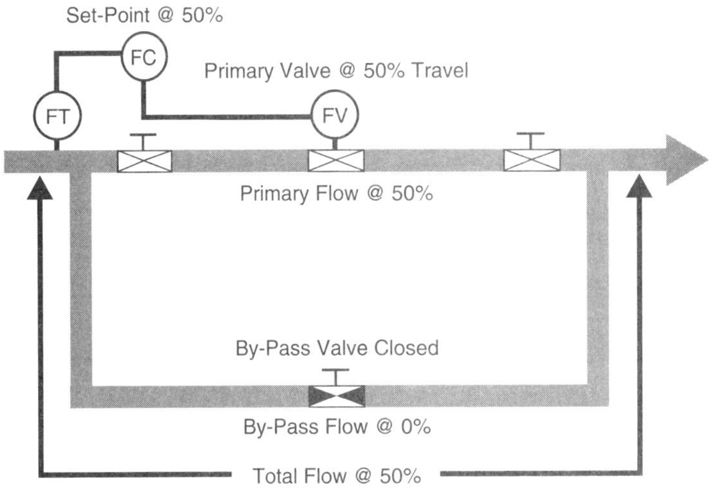

Figure 10.3-1 *Valve in Auto-Control with By-Pass Closed*

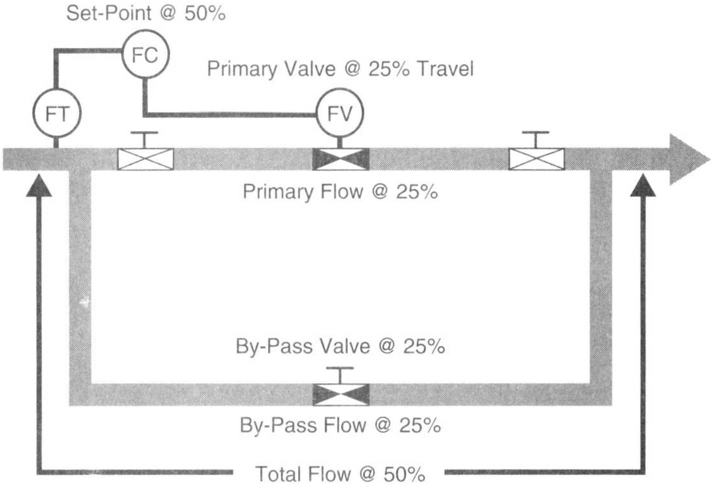

Figure 10.3-2 *Valve in Auto with By-Pass Throttled*

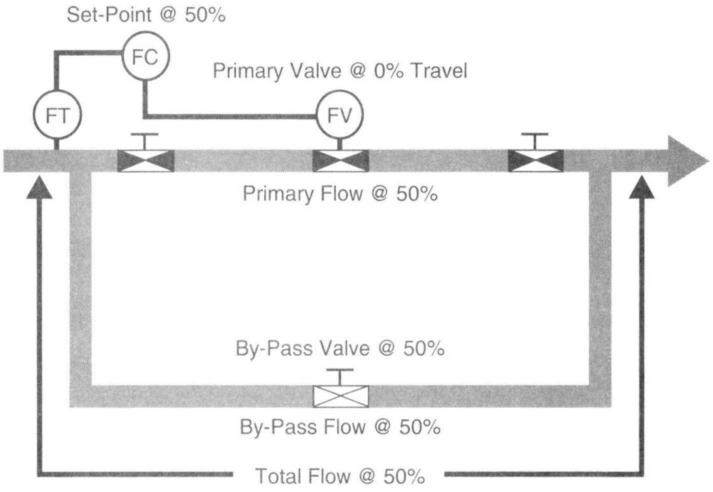

Figure 10.3-3 *Valve Secured/By-Pass Controlling*

10.4 Field Tuning Fundamentals

Optimum tuning of a control loop depends on the particular requirements and demands of the process monitored. Three common types of control response tuning are quarter wave (ultimate period), minimum disturbance, and minimum area. Quarter wave dampening (ultimate period) is the most common desired response. Quarter wave response is characterized by a decaying sinusoidal recovery response signal where each successive cycle attains only one-fourth peak amplitude of the previous cycle. Quarter wave is highly effective at set point recovery, as most disturbances are corrected in four cycles. There are however, some applications where an aggressive disturbance recovery is unacceptable. Minimum disturbance response is sometimes preferred. Minimum disturbance tuning produces high system stability. The process variable remains extremely stable, but recovery time is extended and some set-point offset often remains. Minimum area tuning is similar to quarter wave response. The desired effect of minimum area tuning is to develop a recovery response that permits the minimum deviant sinusoidal area. The minimum sinusoidal area represents an integration of process variable deviation from set-point over a period of time. System response is rapid, but minor oscillations are sometimes present.

Many techniques are employed to restore control loop stability. The first step is to monitor the system to determine baseline conditions. Observe and record minimum/maximum deviation and cycle time, record the as-found controller settings. Compare the phase relationship between the measured variable and the valve position. Reset-induced disturbances typically lag out of phase and the controller output continues to drive the valve past the optimum position. Once the base conditions are determined, judiciously and incrementally slow down the amount of controller reset. Allow several cycles after each adjustment. Compare the measured variable value to the set-point and, if required, correct the offset with the controller set-point control. If the system continues to oscillate with reset adjusted to the slowest value, other adjustments are required.

Observe the phase relationship between the final control element and the process variable. Derivative (rate) induced instability typically leads out of phase, and initially overdrives the valve beyond the optimum position. The magnitude of the step change in relation to the rate of change in the measured variable is derivative induced. Judiciously and incrementally reduce the controller rate adjustment. Allow several cycles after each adjustment.

Gain/proportional band induced instability is generally in phase with the measured variable. Increasing proportional band/decreasing gain increases system stability and decreases system responsiveness. Decreasing proportional band/increasing gain increases system response and reduces system stability. Judiciously and incrementally increase proportion band/decrease gain. Allow several cycles after each adjustment and, if required, correct offset with the controller set-point control. Once the process is stabilized, introduce an upset and monitor the effects. Increase rate to reduce dynamic deviation (not offset) and decrease recovery time. Speed up reset to eliminate any undesired offset (do not exceed cycle time).

Chapter Ten Summary:

- Only trained and qualified technicians should perform tuning and manipulation of active process systems and control loops.
- Before performing any equipment manipulations, establish communications between the board and field operators to review all affected control schemes, develop guidelines, and establish monitoring of the process variable.
- Tank level instruments are often connected to a common standpipe. Manipulating level switch and level control isolation valves can unexpectedly change the stand-pipe level and result in unanticipated alarms and actuations.
- Fully close both isolation valves before opening the vent or drain valves.
- Even with both upper and lower isolation valves closed, full process pressure is still trapped in the pot.
- Use extreme caution when opening the vent and /or drain.
- Manipulating transmitter isolation valves can induce spikes on the sensing lines that could result in an unanticipated alarm or actuation.
- A loss of sensing line reference fluid, or entry of air into sealed systems, creates significant measurement error.
- The transmitter sensor should never be exposed to excessive differential pressure.
- To isolate a transmitter, close the high-side, then open the equalization valve, the close the low-side valve.
- To restore a transmitter, open the equalization valve, and then open the low side valve. Allow sufficient time for the transmitter to equalize, and close the equalization valve. Open the transmitter's high side valve and restore the transmitter to service.
- In auto-control, adjust the manual output to match the auto-output. Once the outputs are equal, transfer to manual-control and ramp the final element to the desired position.
- Manual/auto-transfer manipulates the controller set-point adjustment to match the auto-output signal to the manual demand signal. Once the outputs are equal, place system in auto-control and adjust the controller set-point as required to ramp the final-control element to the desired position.
- To remove a valve from service in automatic, maintain auto-controls and slowly throttle the by-pass valve into service. Observe the primary final control element modulate toward full-closed as the by-pass valve assumes the flow. Set the manual-control output to match the auto-control output and transfer to manual-control.
- To restore a valve to service with the controller in automatic control, manipulate the controller set-point to match the auto-output signal to the manual demand signal. Once the demand signals match, place the controller in automatic. Slowly throttle the by-pass valve closed, and observe the primary final-control element modulate open and into control as the by-pass valve assumes less flow. Adjust the controller set-point to ramp the primary valve to the desired control position.

- To restore a valve to service with the controller in manual-control, adjust the manual-output to ramp the primary valve toward the desired position. Throttle the by-pass closed, and verify that the controller auto-signals respond to process conditions. Set the manual-control output to match the auto-control output and place the controller in auto-control. Adjust the controller set-point to drive the primary valve to the desired position.
- Three common types of control response tuning are quarter wave (ultimate period), minimum disturbance, and minimum area.
- Quarter wave response is characterized by a decaying sinusoidal recovery response signal where each successive cycle attains only one/fourth peak amplitude of the previous cycle. Quarter wave is highly effective at set-point recovery, as most disturbances are corrected in four cycles.
- Minimum disturbance tuning produces high system stability. The process variable remains extremely stable, but recovery time is extended and some set-point offset often remains.
- The desired effect of minimum area tuning is to develop a recovery response that permits the minimum deviant sinusoidal area. System response is rapid, but minor oscillations are sometimes present.
- Reset induced disturbances typically lag "out of phase" and the controller output continues to drive the valve past the optimum position.
- Derivative (rate) induced instability typically leads "out of phase", and initially overdrives the valve beyond the optimum position.
- Gain/proportional band induced instability is generally "in phase" with the measured variable.
- Once the base conditions are determined, judiciously and incrementally slow-down the amount of controller reset. Judiciously and incrementally reduce the controller rate adjustment. Judiciously and incrementally increase proportion band/decrease gain. Once the process is stabilized, introduce an upset and monitor the effects.
- Increase rate to reduce dynamic deviation (not offset) and decrease recovery time.
- Speed up reset to eliminate any undesired offset (do not exceed cycle time).

Chapter Ten Review Questions:

1. List the steps required before performing any equipment manipulations.
2. Compare and contrast common and preferred methods of mounting tank level instruments.
3. State the adverse effects of manipulating tank level instruments mounted on a common stand pipe.
4. State the proper valve sequences to remove and restore level pots to service.
5. State the reason to use caution when venting or draining a level pot.
6. State the adverse effects of manipulating instruments that share common sensing lines.
7. State two sensing line conditions that create significant measurement error.
8. State a condition to avoid when manipulating differential pressure isolation valves.
9. State the normal valve sequences to remove and restore a transmitter to service.
10. State the proper sequence to execute an auto/manual transfer.
11. State the proper sequence to execute a manual/auto transfer.
12. State the proper sequence to remove a valve from service.
13. State the proper sequence to restore a valve to service in auto-control.
14. State the proper sequence to restore a valve to service in manual-control.
15. State the three common types of control response.
16. Describe quarter wave dampening (ultimate period).
17. Describe minimum disturbance tuning.
18. Describe minimum area tuning.
19. Describe the phase relationship between reset-output and valve position.
20. Describe the relationship between derivative-output and valve position.
21. Describe the relationship between proportional-output and valve position.
22. Explain the dynamic control response of reset.
23. Explain the dynamic control response of rate.
24. Explain the dynamic control response of proportional.

Chapter 11

Troubleshooting

OBJECTIVES

Upon completion of Chapter Eleven the student will be able to:

- *Define troubleshooting.*
- *State the expectation of troubleshooting.*
- *List the adverse affects of improper troubleshooting.*
- *Describe the symptom recognition phase of troubleshooting.*
- *State the expectation of the symptom recognition phase of troubleshooting.*
- *List several resources for symptom recognition data.*
- *Describe the symptom analysis phase of troubleshooting.*
- *State the expectation of the symptom analysis phase of troubleshooting.*
- *Define a possible fault.*
- *Define a probable fault.*
- *Describe the system manipulation phase of troubleshooting.*
- *State the expectation of the system manipulation phase of troubleshooting.*
- *List several resources for system manipulation data.*
- *Describe the fault validation phase of troubleshooting.*
- *State the expectation of the fault validation phase of troubleshooting.*
- *List several resources for fault validation data.*
- *Describe the half split system of fault validation.*
- *Describe the process to troubleshoot power supply and grounding faults.*

11.0 Introduction

Troubleshooting is the ordered process of investigating, identifying, and eliminating a fault or combination of faults responsible for undesirable system conditions. Troubleshooting restores a system to desirable operating conditions and eliminates probable replicate failures. Proper troubleshooting techniques are critical to the safe and economical operation of any system. Troubleshooting is not the random replacement of parts until the symptoms of the undesirable condition are cloaked. Eliminating or masking symptoms without correcting the root cause creates a potentially dangerous and expensive situation. An intermittent failure located in a pressure control device allows system pressure to momentarily exceed safe limits. If the recognized symptom is a high-pressure alarm, and the pressure switch that drives the alarm is found damaged, replacement of the pressure switch is obviously required. With the new switch in place, the alarm clears and the system appears to operate properly. An opportunity to identify and correct the real problem is not realized and the hazardous condition continues to exist.

Problem solving requires the logical analysis of collected data. Successful problem solving depends on the proper analysis of complete and correct data. Compare and contrast similar systems with common faults, multiple monitoring points for common data and system conditions before and after an event. Analyze cause and effect relationships to determine potential faults for a specific symptom, and to relate or synchronize multiple symptoms to a single fault. Cause-effect analysis is an important tool in problem solving. It is important to avoid faulty reasoning, such as assuming that since one event preceded another, the two events must be related. Complete data includes current plant documents and maintenance history, equipment manufacture's specification and service updates, user group, trade organization, and governmental publications, and industry events. A systematic approach to problem solving has clearly defined expectations for each step of the evolution.

11.1 Symptom Recognition

The first phase of troubleshooting consists of symptom recognition and analysis. The primary expectation of symptom recognition is to determine the precise validity and identity of the undesirable condition. Symptom recognition is basic data gathering. Identify and record the; immediate and associated alarms or actuations, process conditions, systems testing, and any unusual sights, sounds, or smells at the time of the event. Resources for this data include; first-out alarm recorders, charts and data logs, operator's logs, and on-shift operator interviews. Solicit and record as much information as possible. Once all the conditions are documented, systematically eliminate proven non-relevant conditions. The level control system in Figure 11.1-1 indicates a level-low alarm. Identified symptoms include; low-level alarm in, high-level alarm not in, sight glass indicating low-level, level control valve in the open position, level controller set-point is set at 50% and controller output gauge indicates

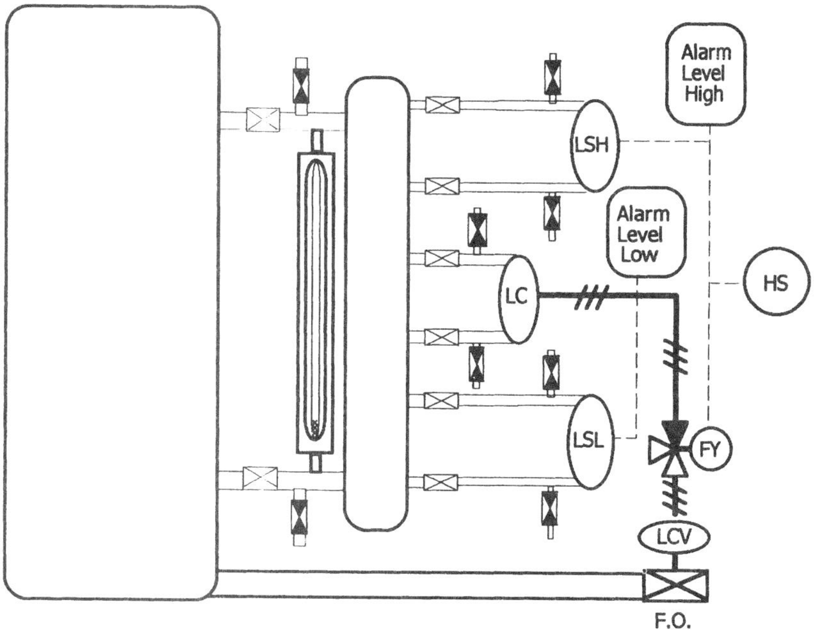

Figure 11.1-1 *Basic Level Control System*

100% output to valve. Instrument air supply valves are open to the controller and valve. The solenoid is emitting a buzzing sound and venting air. The support documentation available is a system logic diagram, Figure 11.1-2 and instrument scale sheets, Figure 11.1-3.

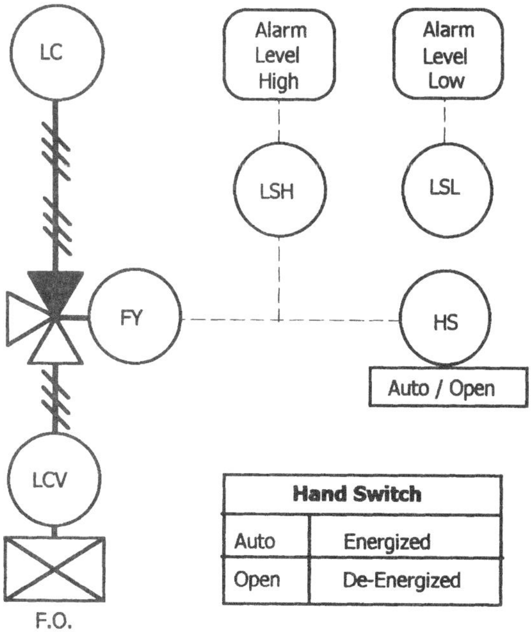

Figure 11.1-2 *Level System Logic Diagram*

LSL Scale Data Sheet	
Trip On Decrease @ 12" H_2O	
Reset	N/A
Contacts	N.C. & Com.
Function	Level Low Alarm

LSH Scale Data Sheet	
Trip On Increase @ 26" H_2O	
Reset	N/A
Contacts	N.C. & Com.
Function	Level High Alarm Fail-Open LCV

LC Scale Data Sheet	
Input	Output
-7" H_2O	3 psig
0" H_2O	9 psig
+7" H_2O	15 psig
Proportional	100% Direct
Set-Point	50%
Reset	Off
Rate	N/A

LCV Scale Data Sheet	
Input	Position
3 psig	0.0 Inches
9 psig	1.5 Inches
15 psig	3.0 Inches
Air Fail-Open	Volt Fail-Open
Benchset Closed	15 psig
Benchset Open	3 psig

Figure 11.1-3 *Level System Scale Sheets*

11.2 Symptom Analysis

Once the symptoms are identified, develop a list or chart and begin the analysis. The expectation of this step is to reduce an inclusive list of possible faults to an exclusive list of probable faults, such as Figure 11.2-1. Symptom analysis begins as an inductive process. For every single symptom, develop an expansive list of possible single point failures. Using a deductive process, review the list of single point failures, and eliminate those that are disproved by reviewing the available information. Also eliminate failures that are not common to all symptoms. The remaining single-point failures that could affect the common components are probable faults. Control valve fault and solenoid fault are identified as probable faults.

11.3 System Manipulation

The third phase is system manipulation. Proper system manipulation often includes recreating known conditions at the time of the event, or judiciously manipulating the system and monitoring component response to controlled changes. It is often helpful to install portable recorders or data loggers to capture real time synchronized data. The expectation of system manipulation is to prove specific elements of a system good and reduce the list of multiple probable faults to a single-point failure. Particular attention is directed to the conditions or

Inductive		Deductive	
Symptom	**Possible Cause**	**Fault Reduction**	**Probable Faults**
Tank Level Low Alarm	Tank Level Low Level Switch Failure	Tank Level Low Alarm Valid	
Sight Glass Level Low	Tank Level Low Level Valve Open	Tank Level Low Alarm Valid	
Valve Open	Hand Switch in Open Controller Output Low Instrument Air Off Valve Failure	Hand Switch in Auto Controller Output 100% Isolation Valves Open	Valve Failure
Controller	Set-PointLow Controller Failure Instrument Air Off	Set-Point 50% Controller Output 100%	
Hand Switch	Hand Switch in Open Hand Switch Failure	Hand Switch in Auto Solenoid Energized (Hot)	
Solenoid	Signal to Open Solenoid Failure	Solenoid Energized (Hot)	Solenoid Failure
Instrument Air	Isolated at Controller Isolated at Valve	Isolation Valves Open	

Figure 11.2-1 *Symptom Recognition and Analysis*

Component	Manipulation	Results	Probable Faults
Tank Level Low Alarm Sight Glass	Manually Close Valve	Level Rises/Alarm Clears	None
Valve	Hand Switch to OPEN Hand Switch to AUTO DriveValve w/Controller Fail/Apply Air Jumper	No Change No Change No Change Valve Strokes w/Remote Air	Undetermined Undetermined Undetermined None
Controller	Lower Set-Point Match Set-Point to Input Raise Set-Point	Controller Output 100% Controller Output 50% Controller Output 0%	None
Hand Switch	Hand Switch in OPEN Hand Switch in AUTO	Solenoid De-Energized Solenoid Energized	None
Solenoid	Energize Solenoid De-Energize Solenoid	No Change No Change	Solenoid Failure
Instrument Air	Check Air at Controller Ccheck Air at Positioner Check Air to Solenoid Check Air from Solenoid	Air Present Air Present Air Present Air Not Present	None None None Solenoid Failure

Figure 11.3-1 *System Manipulation*

components identified in the symptom analysis phase. Figure 11.3-1 indicates a solenoid failure is identified as a probable fault through system manipulation

11.4 Fault Validation

The final phase of troubleshooting is fault validation. The expectation of fault validation is to distinguish and confirm the specific component or condition responsible for the problem identified. Symptom recognition, symptom analysis, and system manipulation lead to the isolation and identification of a specific suspected fault. Fault validation includes the measurement, monitoring, and analysis techniques required to positively identify and documenting the failure. Fault validation requires a working knowledge of the system and access to reliable prints and drawings. A system block diagram is developed that identifies ALL system inputs and ALL system outputs. Each "block" in the system is related to an AND gate. Systematically observe for, and locate, known good conditions. If the output of any "block" is good, then all inputs to that "block" are considered good. If all the inputs to a "block" are good and any of the outputs are bad, then that block is considered faulty. Half-split the system with each observation. Half-splitting is the process of successively dividing a system with each observation until the fault is bracketed. Visualize the signal path traveling through several different modules from left to right. Locating a correct signal at the center of the flow path eliminates all upstream modules, those to the left, as a failure source. The next measurement is taken midway between the remaining suspect modules, those on the right. If the correct signal is not observed, the faulty component is bracketed between the two measurements. The third measurement is taken midway between the first and second. Continue locating known good conditions until the faulty component is identified.

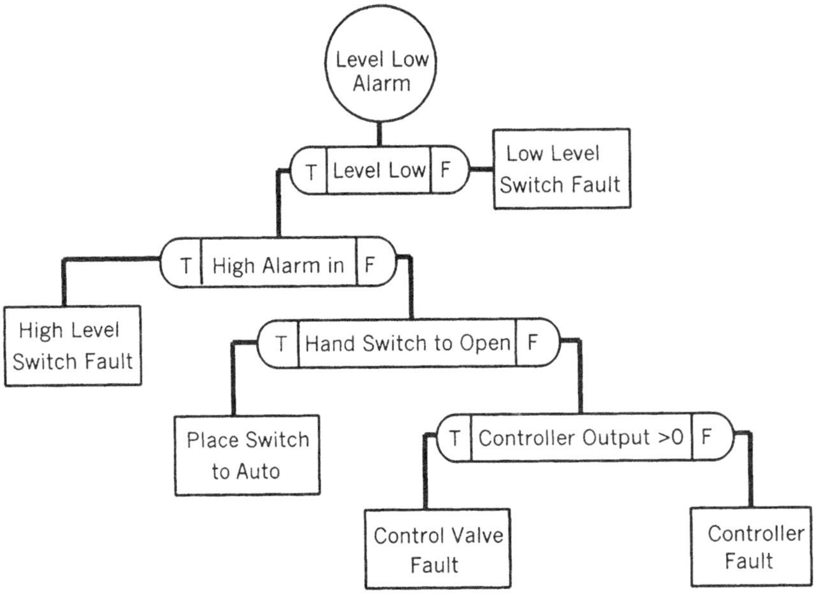

Figure 11.4-1 *High-Order Logic Flow Chart*

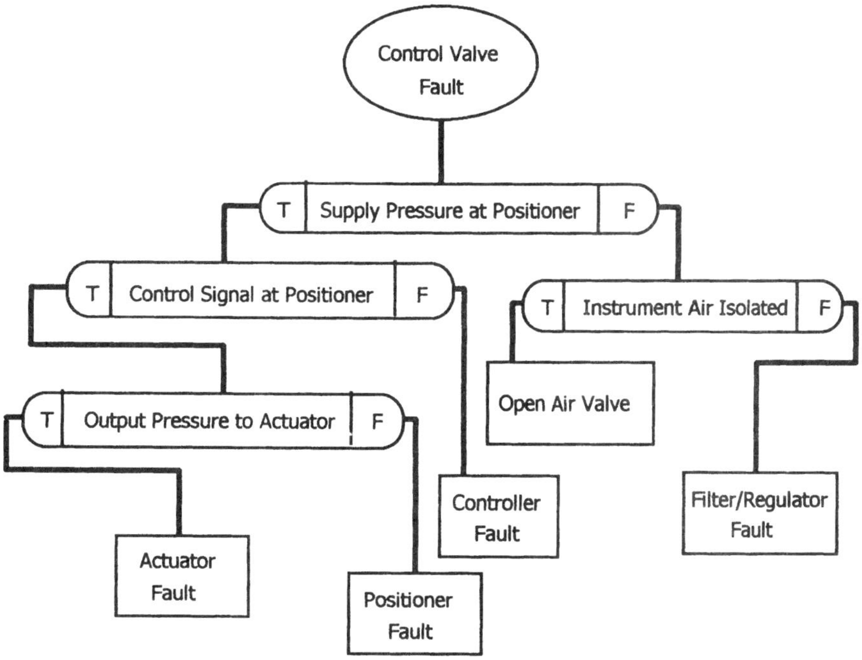

Figure 11.4-2 *Low-Order Logic Flow Chart*

The transition point of a known good condition to an erroneous condition indicates a faulty component. High-order logic flow chart, Figure 11.4-1, isolates the failure to a control valve fault. low-order logic flow chart, Figure 11.4-2, isolates the failure to a controller fault. System manipulation proved the controller good. High-order logic indicated control valve malfunction and low-order logic eliminated all other valve components. The only component between the controller and valve is the solenoid, and system manipulation verified the solenoid control signal changing state at the solenoid. A block diagram is developed of the questioned components, and half-splitting isolates the solenoid as the single-point failure, Figure 11.4-3.

Certain conditions, such as power supply failures, and ground faults present a unique challenge. If the entire system is de-energized, the traditional approaches of symptom recogni-

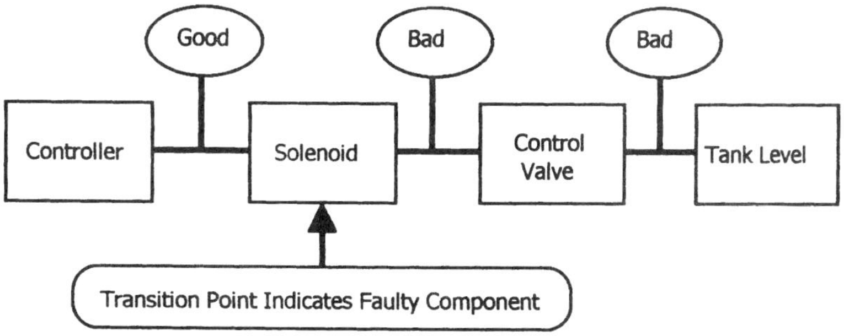

Figure 11.4-3 *Half-Split Bracketing*

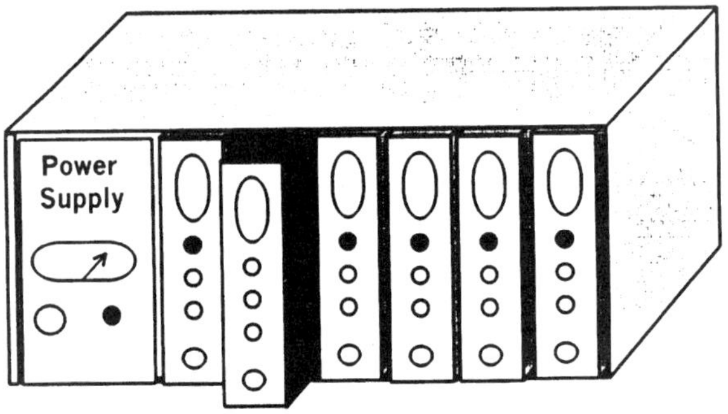

Figure 11.4-4 *Power Supply/Ground Fault Isolation*

tion and system manipulation are impractical. The most direct approach is to unload the bus, restore power, and then systematically restore segments of the system until the ground, or excess load, is detected. Figure 11.4-4 shows a card partially removed from the cabinet's back plane. The faulty component is then replaced and analyzed to validate the root cause of the failure.

Chapter Eleven Summary:

- Troubleshooting is the ordered process of investigating, identifying, and eliminating a fault or combination of faults responsible for undesirable system conditions.
- Troubleshooting restores a system to desirable operating conditions and eliminates probable replicate failures.
- Eliminating or masking symptoms without correcting the root cause creates a potentially dangerous and expensive situation.
- Symptom recognition is basic data gathering.
- The primary expectation of symptom recognition is to determine the precise validity and identity of the undesirable condition.
- Resources for symptom data include; first out alarm recorders, charts and data logs, operator's logs, and on shift operator interviews.
- Symptom analysis develops a list of possible single-point failures for every single symptom, then, using a deductive process, reviews the list and eliminates those that are disproved.
- The expectation of this step is to reduce an inclusive list of possible faults to an exclusive list of probable faults.
- A possible fault is a single-point failure that is capable of generating some of the observed symptoms.
- A probable fault is a single-point failure that is capable of generating all observed symptoms.
- System manipulation is the judicious operation and monitoring of component response to controlled changes.
- The expectation of system manipulation is to prove specific elements of a system good and reduce the list of multiple probable faults to a single-point failure.
- Alarms, portable recorders, and data loggers capture real time synchronized system manipulation data.
- Fault validation is the measurement, monitoring, and analysis techniques used to positively identify and document a failure.
- The expectation of fault validation is to distinguish and confirm the specific component or condition responsible for the problem identified.
- Fault validation requires a working knowledge of the system and access to reliable prints and drawings.
- Half-splitting is the process of successively dividing a system with each observation until the fault is bracketed.
- The most direct approach to troubleshoot power supply and grounding faults is to unload the bus, restore power, and then systematically restore segments of the system until the ground or excess load is detected.

Chapter Eleven Review Questions:

1. Define troubleshooting.
2. State the expectation of troubleshooting.
3. List the adverse affects of improper troubleshooting.
4. List the four phases of troubleshooting.
5. Describe the symptom recognition phase of troubleshooting.
6. State the expectation of the symptom recognition phase of troubleshooting.
7. List several resources for symptom recognition data.
8. State the expectation of the symptom analysis phase of troubleshooting.
9. Describe the difference between a possible fault and a probable fault.
10. Describe the system manipulation phase of troubleshooting.
11. State the expectation of the system manipulation phase of troubleshooting.
12. List several resources for system manipulation data.
13. Describe the fault validation phase of troubleshooting.
14. State the expectation of the fault validation phase of troubleshooting.
15. List several resources for fault validation data.
16. Describe the half-split system of fault validation.
17. Describe the technique to isolate power supply and grounding faults.

Chapter 12

Safety, Certainty, & Professionalism

OBJECTIVES

Upon completion of Chapter Twelve the student will be able to:

- *List potential work area hazards.*
- *List remediation techniques used to reduce area hazards.*
- *List items used to reduce or eliminate environmental risk.*
- *List challenges introduced by safety equipment.*
- *List different approaches used to minimize adverse conditions.*
- *Identify the individual or organization ultimately responsible for your safety.*
- *State the method used to communicate a plan of action to all participants in a team.*
- *State the purpose of a team briefing.*
- *List areas covered in a team briefing.*
- *List actions required prior to initiating an activity.*
- *List techniques used to enhance clear communication.*
- *Differentiate between the terms may, should, and shall.*
- *Differentiate between the terms note, caution, and warning.*
- *State the International Radio Alphabet.*
- *Describe three-part communication.*
- *Define abatement.*
- *Define anomaly.*
- *Define error precursor.*

- *State the impact of disregarding anomalies and error precursors.*
- *Define process restoration.*
- *State the most significant safety contribution operation technicians can make.*
- *State the benefits of the application of professional attitudes and actions.*

12.0 Introduction

Safety awareness is undoubtedly the most significant aspect of professionalism. A momentary lapse in safety awareness can lead to a lifetime of regret. Processing facilities are alive with activity. Conditions are constantly changing. Previously clear walkways are obstructed in a matter of minutes. Risk potentials change unexpectedly. Each work group is naturally focused on their primary task, and can momentarily loose awareness of their surroundings. This can create a dangerous situation. Certain activities are recognized as having high potential to create this condition, and as such are routinely performed only with a dedicated observer. These activities include welding, entering confined spaces, and working in high radiation areas. It is not always practical to have a designated observer assigned to every ongoing task. Much of the time technicians must accept the responsibility to simultaneously perform work and maintain safety awareness.

12.1 Safety

Prior to entering into an area or engaging an activity perform an inspection and evaluation of the surroundings. Note all obvious hazard potentials. These include other work groups, overhead work, slip/trip hazards, and environmental conditions. Resolve these conditions before proceeding. Remediation could include; notifying others in the area of activities, momentarily interrupting overhead work, insuring scaffolds and walks have adequate aprons or skirting in place, requiring the use of lanyards on tools, or securing any items that could become a hazardous falling object. Remove or mark slip/trip hazards. Unavoidable adverse environments may require specialized equipment. Safety harnesses, face shields, SCBA, cooling vest, gas or oxygen monitors, and other equipment can reduce or eliminate the obvious environmental risk, but introduce other considerations. Peripheral vision, hearing, mobility, and stamina or often compromised by the use of safety equipment. Proper training, equipment, and culture can minimize the frequency and extent of adverse conditions, but the ultimate responsibility for safe work activity resides in the individual.

12.2 Certainty

Safe activities are controlled activities. It is critical to maintain control of evolutions at all times. Certainty is a critical component of safety. Prior to any activity develop a plan of action. The plan of action is communicated to all participants in a team briefing. The purpose of the team briefing is to develop and communicate a clear overview of the entire evolution. The briefing should clearly designate specific individuals to perform each task. Key roles identified include; command and control, communications, and support. The brief precisely identifies what equipment is manipulated, what process is affected, and the expected results. Also stated are the specific expectations of the evolution, including the purpose, impact, and relevance. Critical steps and check-points are designated. Potential negative results are identified, and a plan of action to prevent or mitigate potential process transients is discussed. The location of personnel and components involved is established, and a synchronized time line for the initiating event, each critical step, and projected completion is fixed. Prior to initiation of the evolution, perform a physical walk down of the area and equipment. Locate and verify the operability of safety support devices such as fire extinguishers, safety showers, and eyewash stations. Isolate, vent, and drain the system when practical. Notify all at risk personnel in the area or erect exclusion barriers.

During the evolution, use precise communication techniques. Avoid excessive use of acronyms and slang. Carefully select and employ discrete terminology. Ambiguous words

International Radio Alphabet			
Letter	Word	Letter	Word
A	Alpha	N	November
B	Bravo	O	Oscar
C	Charlie	P	Papa
D	Delta	Q	Quebec
E	Echo	R	Romeo
F	Foxtrot	S	Sierra
G	Golf	T	Tango
H	Hotel	U	Uniform
I	India	V	Victor
J	Juliet	W	Whiskey
K	Kilo	X	X-ray
L	Lima	Y	Yankee
M	Mike	Z	Zulu

Figure 12.2-1 *International Radio Alphabet*

and phrases are easily misunderstood. The term "may" delegates permission but does not imply a recommendation. The term "should" implies a recommendation but not a specific requirement. The term "shall" conveys an explicit requirement. The term "note" is a recommendation to take notice, observe, or direct attention to something. The term "caution" is a statement of concern and suggests vigilance. The term "warning" implies that danger is imminent. The International Radio Alphabet, Figure 12.2-1, and three part communications are often used to ensure an accurate exchange of information. Three-part communication consists of the sender stating the original phrase, the receiver restating the original phrase, and the sender acknowledging the restatement as correct. For an example the controller states "take switch Whiskey Zulu Tango to the closed position." The support technician replies "I understand take switch Whiskey Zulu Tango to the closed position." The controller acknowledges with "that is correct". Once the switch is manipulated, the support technician states "switch Whiskey Zulu Tango is in the closed position." The controller repeats back "I understand, switch Whiskey Zulu Tango is in the closed position." The support technician then acknowledges, "that is correct".

12.3 Professionalism

A state of increased awareness is required throughout the entire evolution. All participants must maintain a questioning attitude. An operation technician who has been cycling pumps on the second level for most of the shift may habitually travel back to the second level when requested to cycle a pump located on the third level. Perform a self-check before actuating equipment or manipulating controls. To perform a self-check, momentarily stop and consciously verify your location and the identification of the component. Extremely critical manipulations often require a peer check as well. Communicate identified error-likely situations and error-precursors immediately. Evaluate all anomalies observed, and ensure successful completion of each step before continuing. Error-precursors are often called a near miss or close call. Failure to determine and correct the root cause of a near miss permits the near miss event to persistently re-occur until calamity eventually results.

Document the restoration process. Verify all components and conditions are returned to their desired configuration, and the system left in a safe condition. While no profession is absolutely devoid of all risk, the most significant safety contribution operation technicians can make is to accept ownership and personal accountability for their attitudes and actions. The judicious and dedicated application of professional attitudes and actions creates a safe and productive working environment for everyone.

Chapter Twelve Summary:

- Prior to entering into an area or engaging an activity perform an inspection and evaluation of the surroundings. Note all obvious hazard potentials. Resolve these conditions before proceeding.
- Remediation includes: notifying others in the area, interrupting overhead work, insuring scaffolds and walks have adequate aprons or skirting in place, requiring the use of lanyards on tools, or securing any items that could become a hazardous falling object. Remove or mark slip/trip hazards.
- Safety harnesses, face shields, SCBA, cooling vest, gas or oxygen monitors, and other equipment can reduce or eliminate environmental risk.
- Peripheral vision, hearing, mobility, and stamina or often compromised by the use of safety equipment.
- Proper training, equipment, and culture can minimize the frequency and extent of adverse conditions.
- The ultimate responsibility for safe work activity resides in the individual. Prior to any activity develop a plan of action. Printed procedure/checklists are available for many typical and frequently performed activities. If the activity is fairly complex, and no published guidelines or available, develop a procedure/checklist to provide structure and continuity.
- The plan of action is communicated to all participants in a team briefing.
- The purpose of the team briefing is to develop and communicate a clear overview of the entire evolution.
- The brief precisely identifies equipment manipulated, process affected, and the specific expectations of the evolution. Critical steps and checkpoints are designated, and a plan of action to prevent or mitigate potential process transients is discussed. The location of personnel and components involved is established, and a synchronized time line is fixed.
- Prior to initiation of the evolution perform a physical walk down of the area and equipment. Isolate, vent, and drain the system when practical. Notify all at-risk personnel in the area or erect exclusion barriers.
- Avoid excessive use of acronyms and slang. The International Radio Alphabet and three part communications are often used to ensure an accurate exchange of information.
- Three-part communication consists of the sender stating the original phrase, the receiver restating the original phrase, and the sender acknowledging the restatement as correct.
- Error-precursors are often called a near miss or close call.

- Failure to determine and correct the root cause of a near miss permits the near miss event to persistently occur until calamity eventually results.
- Mitigation is an action that reduces or alleviates the adverse impact of an undesirable event.
- Abatement is an action that stops or eliminates an undesirable condition.
- An anomaly is an event that is irregular and deviant from normal conditions.
- Document the restoration process. Verify all components and conditions are returned to their desired configuration, and the system left in a safe condition.
- The most significant safety contribution operation technicians can make is to accept ownership and personal accountability for their attitudes and actions.
- The judicious and dedicated application of professional attitudes and actions creates a safe and productive working environment for everyone.

Chapter Twelve Review Questions:

1. List potential work area hazards.
2. List remediation techniques used to reduce area hazards.
3. List items used to reduce or eliminate environmental risk.
4. List challenges introduced by safety equipment.
5. List different approaches used to minimize adverse conditions.
6. Identify the individual or organization ultimately responsible for your safety.
7. State the method used to communicate a plan of action to all participants in a team.
8. State the purpose of a team briefing.
9. List areas covered in a team briefing.
10. List actions required prior to initiating an activity.
11. List techniques used to enhance clear communication.
12. State the International Radio Alphabet.
13. Describe three-part communication.
14. Define mitigation.
15. Define abatement.
16. Define anomaly.
17. Define error-precursor.
18. State the impact of disregarding anomalies and error-precursors.
19. Define process restoration.
20. State the most significant safety contribution operation technicians can make.
21. State the benefits of the judicious and dedicated application of professional attitudes and actions.

Index

Note: Page numbers in bold reference figures.